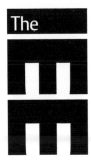

ON-SITE GUIDE
BS 7671:2008

IEE Wiring Regulations 17th Edition

IEE Wiring Regulations Seventeenth Edition
BS 7671:2008 Requirements for Electrical Installations

Published by The Institution of Engineering and Technology, London, United Kingdom

The Institution of Engineering and Technology is registered as a Charity in England & Wales (no. 211014) and Scotland (no. SCO38698).

The Institution of Engineering and Technology is the new institution formed by the joining together of the IEE (The Institution of Electrical Engineers) and the IIE (The Institution of Incorporated Engineers). The new Institution is the inheritor of the IEE brand and all its products and services, such as this one, which we hope you will find useful. The IEE is a registered trademark of the Institution of Engineering and Technology.

© 1992, 1995, 1998, 2002, 2004 The Institution of Electrical Engineers
© 2008 The Institution of Engineering and Technology

First published 1992 (0 85296 537 0)
Reprinted (with amendments) May 1993
Reprinted (with amendments to Appendix 9) July 1993
Reprinted (with amendments) 1994
Revised edition (incorporating Amendment No. 1 to BS 7671:1992) 1995
Reprinted (with new cover) 1996
Revised edition (incorporating Amendment No. 2 to BS 7671:1992) 1998
Second edition (incorporating Amendment No. 1 to BS 7671:2001) 2002 (0 85296 987 2)
Reprinted (with new cover) 2003
Third edition (incorporating Amendment No. 2 to BS 7671:2001) 2004 (0 86341 374 9)
Fourth edition (incorporating BS 7671:2008) 2008 (978-0-86341-854-9)
Reprinted (with amendments) October 2008

This publication is copyright under the Berne Convention and the Universal Copyright Convention. All rights reserved. Apart from any fair dealing for the purposes of research or private study, or criticism or review, as permitted under the Copyright, Designs and Patents Act, 1988, this publication may be reproduced, stored or transmitted, in any form or by any means, only with the prior permission in writing of the publishers, or in the case of reprographic reproduction in accordance with the terms of licences issued by the Copyright Licensing Agency. Enquiries concerning reproduction outside those terms should be sent to the publishers at The Institution of Engineering and Technology, Michael Faraday House, Six Hills Way, Stevenage, SG1 2AY, United Kingdom.

Copies of this publication may be obtained from:
PO Box 96, Stevenage, SG1 2SD, UK
Tel: +44 (0)1438 767328
Email: sales@theiet.org
www.theiet.org/publishing/books/wir-reg/

While the author, publisher and contributors believe that the information and guidance given in this work are correct, all parties must rely upon their own skill and judgement when making use of them. Neither the author nor the publishers assume any liability to anyone for any loss or damage caused by any error or omission in the work, whether such error or omission is the result of negligence or any other cause. Where reference is made to legislation it is not to be considered as legal advice. Any and all such liability is disclaimed.

ISBN 978-0-86341-854-9

Typeset in the UK by The Institution of Engineering and Technology
Printed in the UK by Polestar Wheatons, Exeter

Contents

Cooperating organisations		6
Preface		7
Foreword		9
Section 1	**Introduction**	11
1.1	Scope	11
1.2	The Building Regulations including Part P	12
1.3	Basic information required	13
Section 2	**The service position**	15
2.1	General layout of equipment	15
2.2	Function of components	17
2.3	Separation of gas installation pipework from other services	18
Section 3	**Protection**	19
3.1	Types of protective device	19
3.2	Overload protection	19
3.3	Fault current protection	19
3.4	Protection against electric shock	20
3.5	Automatic disconnection	21
3.6	Residual current devices (RCDs)	21
Section 4	**Earthing and bonding**	27
4.1	Protective earthing	27
4.2	Main protective bonding of metal services	27
4.3	Earthing conductor and main protective bonding conductor cross-sectional areas	28
4.4	Main protective bonding of plastic services	29
4.5	Supplementary equipotential bonding	30
4.6	Additional protection – supplementary equipotential bonding	31
4.7	Supplementary bonding of plastic pipe installations	31
4.8	Earth electrode	31
4.9	Types of earth electrode	31
4.10	Typical earthing arrangements for various types of earthing system	32

Section 5	Isolation and switching	33
5.1	Isolation	33
5.2	Switching off for mechanical maintenance	34
5.3	Emergency switching	34
5.4	Functional switching	35
5.5	Firefighter's switches	35

Section 6	Labelling	37
6.1	Labels to be provided	37

Section 7	Final circuits	43
7.1	Final circuits	43
7.2	Standard final circuits	54
7.3	Installation considerations	59
7.4	Proximity to electrical and other services	61
7.5	Compliance with the Building Regulations	64
7.6	Earthing requirements for the installation of equipment having high protective conductor current	66

Section 8	Locations containing a bath or shower	69
8.1	Summary of requirements	69
8.2	Underfloor heating	72
8.3	Shower cubicle in a room used for other purposes	72

Section 9	Inspection and testing	73
9.1	Inspection and testing	73
9.2	Inspection	73
9.3	Testing	75

Section 10	Guidance on initial testing of installations	77
10.1	Safety and equipment	77
10.2	Sequence of tests	77
10.3	Test procedures	78

Section 11	Operation of RCDs	91
11.1	General test procedure	91
11.2	General purpose RCCBs to BS 4293	91
11.3	General purpose RCCBs to BS EN 61008 or RCBOs to BS EN 61009	92
11.4	RCD protected socket-outlets to BS 7288	92
11.5	Additional protection	92
11.6	Integral test device	92

Appendix 1	Maximum demand and diversity	95
Appendix 2	Maximum permissible measured earth fault loop impedance	99
Appendix 3	Selection of types of cable and flexible cord for particular uses and external influences	105
Appendix 4	Methods of support for cables, conductors and wiring systems	111
Appendix 5	Cable capacities of conduit and trunking	117
Appendix 6	Current-carrying capacities and voltage drop for copper conductors	123
Appendix 7	Certification and reporting	135
Appendix 8	Standard circuit arrangements for household and similar installations	157

8.1 Introduction 157
8.2 Final circuits using socket-outlets complying with BS 1363-2 and fused connection units complying with BS 1363-4 158
8.3 Radial final circuits using 16 A socket-outlets complying with BS EN 60309-2 (BS 4343) 160
8.4 Cooker circuits in household and similar premises 160
8.5 Water and space heating 161
8.6 Height of switches, socket-outlets and controls 161
8.7 Number of socket-outlets 162

Appendix 9	Resistance of copper and aluminium conductors	165
Appendix 10	Selection of devices for isolation and switching	169
Appendix 11	Identification of conductors	171

11.1 Introduction 171
11.2 Addition or alteration to an existing installation 173
11.3 Switch wires in a new installation or an addition or alteration to an existing installation 173
11.4 Intermediate and two-way switch wires in a new installation or an addition or alteration to an existing installation 174
11.5 Line conductors in a new installation or an addition or alteration to an existing installation 174
11.6 Changes to cable core colour identification 174
11.7 Addition or alteration to a d.c. installation 175

Index 176

Errata 179

Cooperating organisations

The IEE acknowledges the contribution made by the following organisations in the preparation of this guide.

Association of Manufacturers of Domestic Appliances
S.A. MacConnacher BSc CEng MIEE

BEAMA Installation Ltd
Eur Ing M.H. Mullins BA CEng FIEE
P. Sayer IEng MIET GCGI

British Cables Association
J.M.R. Haggar BTech(Hons) AMIMMM
C.K. Reed IEng MIET

British Electrotechnical & Allied Manufacturers Association Ltd
P.D. Galbraith IEng MIET MCMI
R.F.B. Lewington MIET

British Standards Institution

City & Guilds of London Institute
H.R. Lovegrove IEng FIET

CORGI
P. Collins MIET

Electrical Contractors' Association
D. Locke IEng MIET ACIBSE
Eur Ing L. Markwell MSc BSc CEng MIET MCIBSE LCGI

Electrical Contractors' Association of Scotland t/a SELECT
D. Millar IEng MIET MILE

ERA Technology Ltd
M.W. Coates BEng

GAMBICA Association Ltd
M. Hadley

Health and Safety Executive
K. Morton BSc CEng MIEE

Institution of Engineering and Technology
M. Coles BEng(Hons) MIEE
G.D. Cronshaw IEng FIET
P.E. Donnachie BSc CEng FIET
J.F. Elliott BSc(Hons) IEng MIEE

Lighting Association
L. Barling
K.R. Kearney IEng MIET

National Inspection Council for Electrical Installation Contracting

Society of Electrical and Mechanical Engineers serving Local Government
C.J. Tanswell CEng MIET MCIBSE

Author
P.R.L. Cook CEng FIEE

Preface

The *On-Site Guide* is one of a number of publications prepared by the IET to provide guidance on certain aspects of BS 7671:2008 *Requirements for Electrical Installations (IEE Wiring Regulations*, 17th Edition). BS 7671 is a joint publication of the British Standards Institution and the Institution of Engineering and Technology.

The scope generally follows that of BS 7671. It includes material not included in BS 7671, provides background to the intentions of BS 7671 and gives other sources of information. However, this guide does not ensure compliance with BS 7671. It is a simple guide to the requirements of BS 7671, and electricians and electrical installers should always consult BS 7671 to satisfy themselves of compliance.

It is expected that persons carrying out work in accordance with this guide will be competent to do so.

Electrical installations in the United Kingdom which comply with the *IEE Wiring Regulations*, BS 7671, should also comply with all relevant Statutory Regulations such as the Electricity at Work Regulations 1989, the Electricity Safety, Quality and Continuity Regulations 2002 and the Building Regulations, in particular Part P. It cannot be guaranteed that BS 7671 complies with all relevant Regulations and it is stressed that it is essential to establish what statutory and other Regulations apply and to install accordingly. For example, an installation in Licensed Premises may have requirements different from or additional to BS 7671 and these will take precedence over BS 7671.

Foreword

This Guide is concerned with limited application of BS 7671 in accordance with paragraph 1.1: Scope.

Part 1

BS 7671 and the *On-Site Guide* are not design guides. It is essential to prepare a schedule of the work to be done prior to commencement or alteration of an electrical installation and to provide all necessary information and operating instructions of any equipment supplied to the user on completion.

Any specification should set out the detailed design and provide sufficient information to enable competent persons to carry out the installation and to commission it.

The specification must provide for all the commissioning procedures that will be required and for the production of any operational manual.

The persons or organisations who may be concerned in the preparation of the specification include:

The Designer(s)
The Installer(s)
The Electricity Distributor
The Installation Owner and/or User
The Architect
The Local Building Control Authority
The Fire Prevention Officer
The CDM Coordinator (the Planning Supervisor)
All Regulatory Authorities
Any Licensing Authority
The Health and Safety Executive.

In producing the specification, advice should be sought from the installation owner and/or user as to the intended use. Often, such as in a speculative building, the detailed intended use is unknown. In those circumstances the specification and/or the operational manual must set out the basis of use for which the installation is suitable.

Precise details of each item of equipment should be obtained from the manufacturer and/or supplier and compliance with appropriate standards confirmed.

The operational manual must include a description of how the system as installed is to operate and all commissioning records. The manual should also include manufacturers' technical data for all items of switchgear, luminaires, accessories, etc. and any special instructions that may be needed. The Health and Safety at Work etc. Act 1974 Section 6 and the Construction (Design and Management) Regulations 2007 are concerned with the provision of information. Guidance on the preparation of technical manuals is given in BS 4884 (Specification for technical manuals) and BS 4940 (Building and civil engineering). The size and complexity of the installation will dictate the nature and extent of the manual.

Introduction 1

1.1 Scope

This Guide is for electricians (for simplicity, the term electrician has been used for electricians and electrical installers). It covers the following installations:

a domestic installations generally, including off-peak supplies, and supplies to associated garages, outbuildings and the like

b industrial and commercial single- and three-phase installations where the distribution board(s) or consumer unit is located at or near the distributor's cut-out.

Note: Special Installations or Locations (Part 7 of BS 7671) are generally excluded from this Guide. Advice is given on installations in locations containing a bath or shower (Section 8).

Part 7

This Guide is restricted to installations:

313

i at a supply frequency of 50 hertz
ii at a nominal voltage of 230 V a.c. single-phase or 230/400 V a.c. three-phase
iii fed through a distributor's cut-out having a fuse or fuses to BS 1361 Type II or through fuses to BS 88-2 or BS 88-6
iv with a maximum value of the earth fault loop impedance outside the consumer's installation as follows:

▶ TN-C-S system (earth return via combined neutral and earth conductor): 0.35 Ω, Figure 2.1
▶ TN-S system (earth return via separate earth conductor): 0.8 Ω, Figure 2.2
▶ TT system (earth return via consumer's earth only): 21 Ω excluding consumer's earth electrode, Figure 2.3

Note: 21 Ω is the stated maximum resistance of the distributor's earth electrode at the supply transformer. The resistance of the consumer's installation earth electrode should be as low as practicable. A value exceeding 200 Ω may not be stable. Refer to Table 41.5, note 2 and Regulation 542.2.2 of BS 7671.

This Guide contains information which may be required in general installation work, e.g. conduit and trunking capacities, bending radii of cables, etc.

The Guide introduces the use of standard circuits, which are discussed in Section 7, however, because of simplification this Guide may not give the most economical result.

This Guide is not a replacement for BS 7671, which should always be consulted. Defined terms according to Part 2 of BS 7671 are used.

In compliance with the definitions of BS 7671, throughout this Guide the term line conductor is used instead of phase conductor and 'live part' is used to refer to a conductor or conductive part intended to be energised in normal use, including a neutral conductor. The terminals of electrical equipment are identified by the letters L, N and E (or PE).

Further information is available in the series of Guidance Notes published by the Institution.

1.2 The Building Regulations including Part P

Note: Approved Documents and guidance can be freely downloaded from the Department for Communities and Local Government (DCLG) website: www.planningportal.gov.uk

Persons carrying out electrical work in dwellings must comply with the Building Regulations of England and Wales, in particular Part P (Electrical safety – dwellings).

The Building Regulations do not apply in Scotland or Northern Ireland. In Scotland the requirements of the Building Regulations (Scotland) 2004 apply, in particular Regulation 9, and in Northern Ireland the Building Regulations (Northern Ireland) 2000 (as amended) apply.

Persons responsible for work within the scope of Part P of the Building Regulations may also be responsible for ensuring compliance with other Parts of the Building Regulations, where relevant, particularly if there are no other parties involved with the work. Building Regulations requirements relevant to electricians carrying out electrical work include:

Part A (Structure): depth of chases in walls, and size of holes and notches in floor and roof joists;
Part B (Fire safety): fire safety of certain electrical installations; provision of fire alarm and fire detection systems; fire resistance of penetrations through floors and walls;
Part C (Site preparation and resistance to moisture): moisture resistance of cable penetrations through external walls;
Part E (Resistance to the passage of sound): penetrations through floors and walls;
Part F (Ventilation): ventilation rates for dwellings;
Part L (Conservation of fuel and power): energy efficient lighting;
Part M (Access to and use of buildings): heights of switches, socket-outlets and consumer units;
Part P (Electrical safety – dwellings).

Guidance for electricians on the Building Regulations, including all the parts above is given in the IEE publication *Electrician's Guide to the Building Regulations*.

1.3 Basic information required

313.1

Before starting work on an installation that requires a new supply, the electrician should obtain the following information from the distributor:

- i the number of phases to be provided
- ii the distributor's requirement for cross-sectional area and length of meter tails
- iii the maximum prospective fault current (pfc) at the supply terminals
- iv the maximum earth fault loop impedance (Z_e) of the earth fault path outside the consumer's installation
- v the type and rating of the distributor's fusible cut-out or protective device
- vi the distributor's requirement regarding the size of main protective bonding conductors   544.1
- vii the earthing arrangement and type of system   312.3
- viii the arrangements for the incoming cable and metering.

For existing installations, electricians should satisfy themselves as to the suitability of the supply including the earthing arrangement.

The service position 2

2.1 General layout of equipment

The general layout of the equipment at the service position is shown in Figures 2.1, 2.2 and 2.3.

▼ **Figure 2.1**   TN-C-S system (PME supply) meter position arrangement

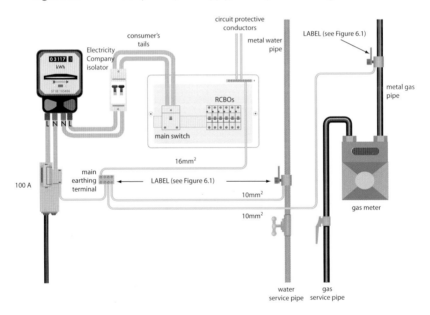

Note: An isolator is often not installed by the distributor.

▼ **Figure 2.2** TN-S system (cable sheath earth) meter position arrangement

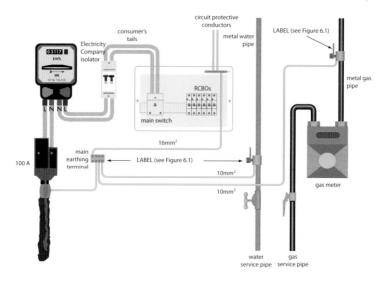

Note: An isolator is often not installed by the distributor.

▼ **Figure 2.3** TT system (no distributor's earth) meter position arrangement

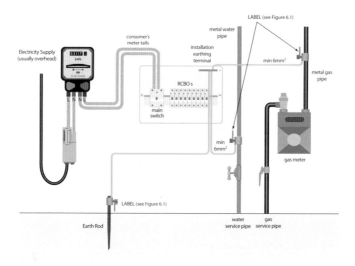

Note: An isolator is often not installed by the distributor. See Table 4.2 for sizing of earthing conductor of TT systems.

2.2 Function of components

2.2.1 Distributor's cut-out
This will be sealed to prevent the fuse being withdrawn by unauthorised persons. When the meter tails and consumer unit are installed in accordance with the requirements of the distributor, the cut-out may be assumed to provide fault current protection up to the consumer's main switch.

2.2.2 Electricity meter
This will be sealed by the meter owner to prevent interference by unauthorised persons.

2.2.3 Meter tails
These are part of the consumer's installation. They should be insulated and sheathed or insulated and enclosed in conduit or trunking. They are provided by the installer. 521.10.1

Polarity should be indicated by the colour of the insulation and the minimum cable size should be 25 mm². The distributor may specify the maximum length and the minimum cross-sectional area (see 1.3). 514.3

Where the meter tails are protected against fault current by the distributor's cut-out the method of installation, maximum length and minimum cross-sectional area must comply with the requirements of the distributor. 434.3(iv)

2.2.4 Supplier's switch
Some suppliers may provide and install a suitable switch between the meter and the consumer unit. This permits the supply to the installation to be interrupted without withdrawing the distributor's fuse in the cut-out.

2.2.5 Consumer's controlgear
A consumer unit (to BS EN 60439-3 Annex ZA) is for use on single-phase installations up to 100 A. It includes: 530.3.4

- a double-pole isolator,
- fuses, circuit-breakers or RCBOs for protection against overload and fault currents, and
- RCDs for additional protection against electric shock.

Alternatively, a separate main switch and distribution board may be provided.

2.3 Separation of gas installation pipework from other services

Gas installation pipes must be spaced:

a at least 150 mm away from electricity meters, controls, electrical switches or sockets, distribution boards or consumer units;
b at least 25 mm away from electricity cables.

(BS 6891:2005 *Installation of low pressure gas pipework in domestic premises*, clause 8.16.2)

▼ **Figure 2.4**   Separation from gas pipes and meters

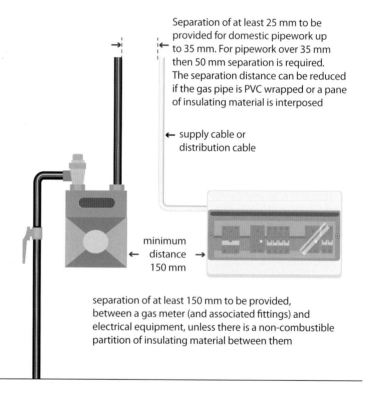

Separation of at least 25 mm to be provided for domestic pipework up to 35 mm. For pipework over 35 mm then 50 mm separation is required. The separation distance can be reduced if the gas pipe is PVC wrapped or a pane of insulating material is interposed

← supply cable or distribution cable

minimum
← distance →
150 mm

separation of at least 150 mm to be provided, between a gas meter (and associated fittings) and electrical equipment, unless there is a non-combustible partition of insulating material between them

Protection   3

3.1 Types of protective device

The consumer unit (or distribution board) contains devices for the protection of the final circuits against:

    i   overload   433
    ii short-circuit   434
    iii earth fault.   434

Functions **i** and **ii** are carried out usually by one device, i.e. a fuse or circuit-breaker.

Function **iii** may be carried out by the fuse or circuit-breaker provided for functions **i** and 434
ii or by an RCD.   411

An RCBO, being a combined circuit-breaker and RCD, will carry out functions **i**, **ii** and **iii**.

3.2 Overload protection   Appx 3

Overload protection is given by the following devices:

- fuses to BS 88-2 or BS 88-6; BS 1361 and BS 3036;
- miniature circuit-breakers to BS 3871-1 Types 1, 2 and 3;
- circuit-breakers to BS EN 60898 types B, C and D; and
- residual current circuit-breakers with integral overcurrent protection (RCBOs) to BS EN 61009-1.

3.3 Fault current protection

When a consumer unit to BS EN 60439-3 or BS 5486:Part 13, or a fuseboard having fuselinks to BS 88-2 or BS 88-6 or BS 1361 is used, then fault current protection will be given by the overload protective device.

For other protective devices the breaking capacity must be adequate for the prospective fault current at that point.

3.4 Protection against electric shock

3.4.1 Basic protection

Electrical insulation and enclosures and barriers give protection against contact with live parts. Non-sheathed insulated conductors must be protected by conduit or trunking or be within a suitable enclosure.

A 30 mA RCD may be provided to give additional protection against contact with live parts but must not be relied upon for primary protection.

3.4.2 Fault protection

Fault protection is given by limiting the magnitude and duration of voltages that may appear under earth fault conditions between simultaneously accessible exposed-conductive-parts of equipment, and between them and extraneous-conductive-parts or earth. This may be effected by:

a connecting all exposed-conductive-parts to the main earthing terminal via circuit protective conductors, and selecting appropriate fault current protective devices (fuses, circuit-breakers, MCCBs or RCDs) that will operate in the event of a fault, or

b the use of double or reinforced insulation.

3.4.3 SELV and PELV

SELV

Separated extra-low voltage (SELV) systems:

a are supplied from isolated safety sources such as a safety isolating transformer to BS EN 61558-2-6
b have no live part connected to earth or the protective conductor of another system
c have basic insulation from other SELV and PELV circuits
d have double or reinforced insulation or basic insulation plus earthed metallic screening from LV circuits
e have no exposed-conductive-parts connected to earth, to exposed-conductive-parts or protective conductors of another circuit.

PELV

Protective extra-low voltage (PELV) systems must meet all the requirements for SELV, except that the circuits are not electrically separated from earth.

For SELV and PELV systems basic protection need not be provided if voltages do not exceed the following:

Location	SELV	PELV
Dry areas	25 V a.c. or 60 V d.c	25 V a.c. or 60 V d.c
Immersed equipment	Protection required at all voltages	Protection required at all voltages
Locations containing a bath or shower, swimming pools, saunas	Protection required at all voltages	Protection required at all voltages
Other areas	12 V a.c. or 30 V d.c.	12 V a.c. or 30 V d.c.

3.5 Automatic disconnection

411

3.5.1 Standard circuits

For the standard final circuits given in Section 7, the correct disconnection time is obtained for the protective devices by limiting the maximum circuit lengths.

3.5.2 Disconnection times – TN circuits

Table 41.1

A disconnection time of not more than 0.4 s is required for final circuits with a rating (I_n) not exceeding 32 A.

411.3.2.2

A disconnection time of not more than 5 s is required for

411.3.2.3

- final circuits exceeding 32 A, and
- distribution circuits.

3.5.3 Disconnection times – TT circuits

Table 41.1

The required disconnection times for TT systems can, except in the most exceptional circumstances outside the scope of this guide, only be achieved by protecting every circuit with an RCD.

411.3.2.4
411.5.3

3.6 Residual current devices (RCDs)

Note: Residual current device (RCD) is a device type that includes residual current circuit-breakers (RCCBs), residual current circuit-breakers with integral overcurrent protection (RCBOs) and socket-outlets incorporating RCDs (SRCDs).

3.6.1 Protection by an RCD

RCDs are required:

i where the earth fault loop impedance is too high to provide the required disconnection, e.g. where the distributor does not provide a connection to the means of earthing – TT system

411.5

ii for socket-outlet circuits in domestic and similar installations

411.3.3

iii for circuits of locations containing a bath or shower

701.411.3.3

411.3.3	**iv**	for circuits supplying mobile equipment not exceeding 32 A for use outdoors
522.6.7	**v**	for cables without earthed metallic covering installed in walls or partitions at a depth of less than 50 mm and not protected by earthed steel conduit or similar
522.6.8	**vi**	for cables without earthed metallic covering installed in walls or partitions with metal parts (not including screws or nails) and not protected by earthed steel conduit or the like.

Note: Cables installed on the surface do not require RCD protection.

30 mA RCDs are required for **ii** to **vi** above.

RCDs may be omitted for:

411.3.3   **a**   specific labelled sockets, such as a socket for a freezer. However, the circuit cables must not require RCDs as per **v** and **vi** above, that is, circuit cables must be enclosed in earthed steel conduit or have an earthed metal sheath or be at a depth of 50mm in a wall or partition without metal parts.

           b   socket-outlet circuits in industrial and commercial premises where the use of equipment and work on the building fabric and electrical installation is controlled by skilled or instructed persons.

3.6.2 Applications of RCDs

314.1 Installations are required to be divided into circuits to avoid hazards and minimize inconvenience in the event of a fault and to take account of danger that might arise from the failure of a single circuit, e.g. a lighting circuit.

a   TN conduit installations

Where cables in walls or partitions have an earthed metallic covering or are installed in steel conduit or similar, 30 mA RCD protection is still required in the following cases:

- circuits of locations containing a bath or shower,
- circuits with socket-outlets not exceeding 20 A,
- mobile equipment not exceeding 32 A for use outdoors, and
- the arrangement in Figure 3.1.

▼ **Figure 3.1** Typical split consumer unit with one 30 mA RCD, suitable for TN installations with cables in walls or partitions having an earthed metallic covering or enclosed in earthed steel conduit or the like

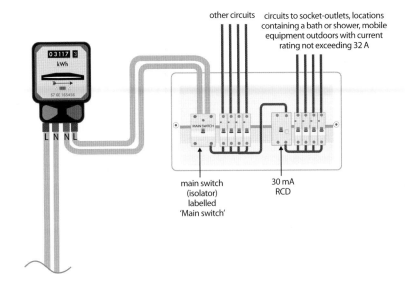

b TT conduit installations

For TT installations, all circuits must be RCD protected. If cables in walls or partitions have an earthed metallic covering or are installed in earthed steel conduit, 30 mA RCDs will be required for:

- ▶ circuits of locations containing a bath or shower,
- ▶ circuits with socket-outlets not exceeding 20 A,
- ▶ mobile equipment not exceeding 32 A for use outdoors.

The rest of the installation needs protecting by a 100 mA RCD (see Figure 3.2).

▼ **Figure 3.2** Typical split consumer unit with time-delayed RCD as main switch, suitable for TT and TN installations with cables in walls or partitions having an earthed metallic covering or enclosed in earthed steel conduit or the like

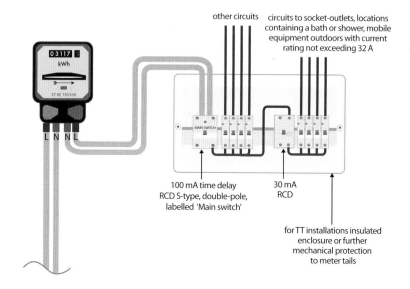

For TT installations with cables installed in walls or partitions having no earthed metallic covering or not installed in earthed conduit or the like, protection by 30 mA RCDs will be required for all circuits, see Figures 3.3 and 3.4.

The enclosures of RCDs or consumer units incorporating RCDs in TT installations should have an all-insulated or Class II construction, or additional precautions recommended by the manufacturer need to be taken to prevent faults to earth on the supply side of the 100 mA RCD.

c RCBOs

The use of RCBOs, see Figure 3.3, will minimize inconvenience in the event of a fault and is applicable to all systems.

Such a consumer unit arrangement also easily allows individual circuits, such as to specifically labelled sockets or fire alarms, to be protected by a circuit-breaker without RCD protection. Such circuits will usually need to be installed in earthed metal conduit or wired with earthed metal-sheathed cables.

▼ **Figure 3.3**   Consumer unit with RCBOs, suitable for all installations (TN and TT)

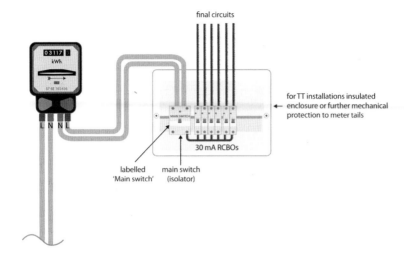

d   Split board with two 30 mA RCDs

The division of an installation into two parts with separate 30 mA RCDs will ensure that part of the installation will remain on supply in the event of a fault, see Figure 3.4.

▼ **Figure 3.4**   Split consumer unit with separate main switch and two 30 mA RCDs

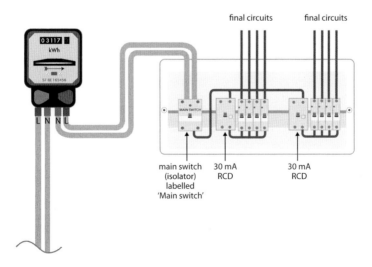

e Three-way split board with two 30 mA RCDs

The three-way division of an installation to provide ways unprotected by RCDs for, say, fire systems and for two separate 30 mA RCDs to ensure that part of the installation will remain on supply in the event of a fault. Unprotected circuits will usually need to be installed in earthed metal conduit or wired with earthed metal-sheathed cables, see Figure 3.5.

▼ **Figure 3.5** Three-way split consumer unit with separate main switch, two 30 mA RCDs and circuits without RCD protection

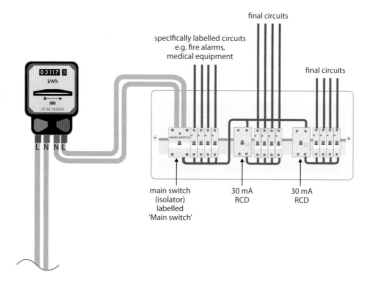

Earthing and bonding   4

4.1 Protective earthing

The purpose of protective earthing is to ensure that, in the event of a fault (line conductor to exposed-conductive-part), sufficient current flows to operate the protective device (fuse to blow, circuit-breaker to trip, RCD to trip) in the required time.

Every exposed-conductive-part (a conductive part of equipment that can be touched and which is not a live part but which may become live when basic insulation fails) shall be connected by a protective conductor to the main earthing terminal.   411.4.2
411.5.1

4.2 Main protective bonding of metal services

(Figures 2.1, 2.2, 2.3)

The purpose of protective equipotential bonding is to reduce the voltages between the various exposed-conductive-parts and extraneous-conductive-parts of an installation, during a fault to earth and in the event of a fault on the distributor's network.

Main protective bonding conductors are required to connect the following metallic parts to   411.3.1.2
the main earthing terminal, where they are extraneous-conductive-parts*:

i metal water installation pipes
ii metal gas installation pipes
iii other metal installation pipes (including oil and gas supply pipes) and ducting
iv metal central heating and air conditioning systems
v exposed metallic structural parts of the building
vi lightning protection systems (where required by BS EN 62305).

* Extraneous-conductive-part: a conductive part such as a metal pipe, liable to introduce earth potential into the building.

4.3 Earthing conductor and main protective bonding conductor cross-sectional areas

The minimum cross-sectional areas (csa) of the earthing conductor and main protective bonding conductors are given in Table 4.1. For TT supplies, refer to Table 4.2.

▼ **Table 4.1** Earthing conductor and main protective bonding conductor sizes (copper equivalent) for TN-S and TN-C-S supplies

	Line conductor or neutral conductor of PME supplies	mm^2	4	6	10	16	25	35	50	70
542.3 543.1	Earthing conductor not buried or buried protected against corrosion and mechanical damage – see notes	mm^2	6	6	10	16	16	16	25	35
544.1.1	Main protective bonding conductor – see notes	mm^2	6	6	6	10	10	10	16	25
Table 54.8	Main protective bonding conductor for PME supplies (TN-C-S)	mm^2	10	10	10	10	10	10	16	25

Notes:

543.2.3   **1**   Protective conductors (including earthing and bonding conductors) of 10 mm^2 cross-sectional area or less shall be copper.

2   The distributor may require a minimum size of earthing conductor at the origin of the supply of 16 mm^2 copper or greater for TN-S and TN-C-S supplies.

542.3.1
Table 54.1   **3**   Buried earthing conductors must be at least:
- 25 mm^2 copper if not protected against mechanical damage or corrosion
- 50 mm^2 steel if not protected against mechanical damage or corrosion
- 16 mm^2 copper if not protected against mechanical damage but protected against corrosion
- 16 mm^2 coated steel if not protected against mechanical damage but protected against corrosion.

4   The distributor should be consulted when in doubt.

▼ **Table 4.2** Copper earthing conductor cross-sectional area (csa) for TT supplies for earth fault loop impedances not less than 1 ohm

Buried			Not buried		
Unprotected	Protected against corrosion	Protected against corrosion and mechanical damage	Unprotected	Protected against corrosion	Protected against corrosion and mechanical damage
mm²	mm²	mm²	mm²	mm²	mm²
25	16	2.5	4	4	2.5

Notes:
1. Protected against corrosion by a sheath.
2. For impedances less than 1 ohm determine as per Regulation 543.1.2.
3. The main protective bonding conductors shall have a cross-sectional area of not less than half that required for the earthing conductor and not less than 6 mm² (Regulation 544.1.1).

Note that:

    i only copper conductors should be used; copper covered aluminium conductors or aluminium conductors or structural steel can only be used if special precautions outside the scope of this Guide are taken   543.2.3

    ii bonding connections to incoming metal services should be made as near as practicable to the point of entry of the services into the premises, but on the consumer's side of any insulating section   544.1.2

    iii where practicable the connection to the gas, water, oil, etc., service should be within 600 mm of the service meter, or at the point of entry to the building if the service meter is external and must be on the consumer's side before any branch pipework and after any insulating section in the service. The connection must be made to hard pipework, not to soft or flexible meter connections   544.1.2

    iv the connection must be made using clamps (to BS 951) and be suitably protected against corrosion at the point of contact   542.3.2

    v if incoming gas or water services are of plastic, main bonding connections are to be made to metal installation pipes only, where required.

4.4 Main protective bonding of plastic services

There is no requirement to main bond an incoming service where the incoming service pipe and the pipework within the installation are both of plastic. Where there is a plastic incoming service and a metal installation within the premises, main bonding is recommended unless it has been confirmed that any metal pipework within the building is not introducing earth potential. All bonding connections are to be applied to the consumer's side of any meter, main stopcock or insulating insert.

4.5 Supplementary equipotential bonding

The purpose of supplementary equipotential bonding is to reduce the voltage between the various exposed-conductive-parts and extraneous-conductive-parts of a location during a fault to earth.

411.3.2.6 Where a required disconnection time cannot be achieved, supplementary bonding shall be applied.

Note: Disconnection must be still be achieved in the event of a fault.

The cross-sectional area of supplementary bonding conductors is given in Table 4.3.

▼ **Table 4.3** Supplementary bonding conductors

Size of circuit protective conductor (mm²)	Minimum cross-sectional area of supplementary bonding conductor (mm²)					
	Exposed-conductive-part to extraneous-conductive-part		Exposed-conductive-part to exposed-conductive-part		Extraneous-conductive-part to extraneous-conductive-part*	
	mechanically protected	not mechanically protected	mechanically protected	not mechanically protected	mechanically protected	not mechanically protected
1	2	3	4	5	6	
1.0	1.0	4.0	1.0	4.0	2.5	4.0
1.5	1.0	4.0	1.5	4.0	2.5	4.0
2.5	1.5	4.0	2.5	4.0	2.5	4.0
4.0	2.5	4.0	4.0	4.0	2.5	4.0
6.0	4.0	4.0	6.0	6.0	2.5	4.0
10.0	6.0	6.0	10.0	10.0	2.5	4.0
16.0	10.0	10.0	16.0	16.0	2.5	4.0

* If one of the extraneous-conductive-parts is connected to an exposed-conductive-part, the bond must be no smaller than that required for bonds between exposed-conductive-parts – column 3 or 4.

4.6 Additional protection – supplementary equipotential bonding

415.2

Supplementary equipotential bonding is required in some of the locations and installations of Part 7 of BS 7671.

If the installation meets the requirements for earthing and bonding, there is no specific requirement in BS 7671 for supplementary equipotential bonding of:

- kitchen pipes, sinks or draining boards
- metallic boiler pipework
- metal furniture in kitchens
- metallic pipes to wash hand basins and WCs
- locations containing a bath or shower, providing the requirements of 701.415.2 are met.

Note: Metallic waste pipes in contact with Earth must be bonded to the main earthing terminal as they are extraneous-conductive-parts.

4.7 Supplementary bonding of plastic pipe installations

Supplementary bonding is not required to metallic parts supplied by plastic pipes.

4.8 Earth electrode (Figure 2.3)

This is connected to the main earthing terminal by the earthing conductor and provides part of the earth fault loop for a TT installation.

542.1.4

It is recommended that the earth fault loop impedance for TT installations does not exceed 200 ohms.

Table 41.5, Note 2

Metallic gas or water utility or other metallic service pipes are not to be used as the earth electrode, although they must be bonded as paragraph 4.2.

542.2.4

Note: Regulation 542.2.4 permits the use of privately owned water supply pipework for use as an earth electrode where precautions are taken against its removal and it has been considered for such use. This relaxation will not apply to a domestic installation.

4.9 Types of earth electrode

542.2.1

The following types of earth electrode are recognised:

- i   earth rods or pipes
- ii   earth tapes or wires
- iii earth plates
- iv   underground structural metalwork embedded in foundations

v welded metal reinforcement of concrete embedded in the Earth (excluding pre-stressed concrete)

vi lead sheaths and metal coverings of cables, which must meet the following conditions:

    a the sheath or covering shall be in effective contact with Earth,

    b the consent of the owner of the cable shall be obtained, and

    c arrangements shall be made for the owner of the cable to warn the owner of the electrical installation of any proposed change to the cable or its method of installation which might affect its suitability as an earth electrode.

4.10 Typical earthing arrangements for various types of earthing system

Figures 2.1, 2.2 and 2.3 show single-phase arrangements and three-phase arrangements are similar.

The protective conductor sizes shown in the above-mentioned figures refer to copper conductors and are related to 25 mm² supply tails from the meter.

For TT systems protected by an RCD with an earth electrode resistance 1 ohm or greater, the earthing conductor size need not exceed 2.5 mm² if protected against corrosion by a sheath and if also protected against mechanical damage; otherwise, see Table 4.2.

The earthing bar is sometimes used as the main earthing terminal; however, means must be provided in an accessible position for disconnecting the earthing conductor to facilitate testing of the earthing.

Note: For TN-S and TN-C-S installations, advice about the availability of an earthing facility and the precise arrangements for connection should be obtained from the distributor or supplier.

Isolation and switching   5

5.1 Isolation
537.1
537.2

5.1.1 Requirement

Means of isolation should be provided:
132.15.1

i at the origin of the installation
537.1.4

A main linked switch or circuit-breaker should be provided as a means of isolation and of interrupting the supply on load.

For single-phase household and similar supplies that may be operated by unskilled persons, a double-pole device must be used for both TT and TN systems.

For a three-phase supply to an installation forming part of a TT system, an isolator must interrupt the line and neutral conductors. In a TN-S or TN-C-S system only the line conductors need be interrupted.

ii for every circuit
537.2.1.1

Other than at the origin of the installation, every circuit or group of circuits that may have to be isolated without interrupting the supply to other circuits should be provided with its own isolating device. The device must switch all live conductors in a TT system and all line conductors in a TN system.

iii for every item of equipment
537.2.1.2

iv for every motor

Every fixed electric motor should be provided with a readily accessible and easily operated device to switch off the motor and all associated equipment including any automatic circuit-breaker. The device must be so placed as to prevent danger.
132.15.2

5.1.2 The switchgear

The position of the contacts of the isolator must either be externally visible or be clearly, positively and reliably indicated.
537.2.2.2

The device must be designed or installed to prevent unintentional or inadvertent closure.
537.2.2.3

5

537.2.2.6 Each device used for isolation must be clearly identified by position or durable marking to indicate the installation or circuit that it isolates.

537.2.2.5 If it is installed remotely from the equipment to be isolated, the device must be capable of being secured in the OPEN position.

Guidance on the selection of devices for isolation is given in Appendix 10.

537.3 5.2 Switching off for mechanical maintenance

537.3.1.1 A means of switching off for mechanical maintenance is required where mechanical maintenance may involve a risk of injury – for example, from mechanical movement of machinery or hot items when replacing lamps.

537.3.1.2 The means of switching off for mechanical maintenance must be able to be made secure to prevent electrically powered equipment from becoming unintentionally started during the mechanical maintenance, unless the means of switching off is continuously under the control of the person performing the maintenance.

Each device for switching off for mechanical maintenance must:

537.3.2.1	i	where practicable, be inserted in the main supply circuit
537.3.2.5	ii	be capable of switching the full load current
537.3.2.2	iii	be manually operated
537.3.2.2	iv	have either an externally visible contact gap or a clearly and reliably indicated OFF position. An indicator light should not be relied upon.
537.3.2.3	v	be selected and installed so as to prevent inadvertent or unintentional switching on
537.3.2.4	vi	be installed and durably marked so as to be accessible and readily identifiable.

537.3.2.6 A plug and socket-outlet or similar device of rating not exceeding 16 A may be used for switching off for mechanical maintenance.

537.4 5.3 Emergency switching

537.4.1.1 An emergency switch is to be provided for every part of an installation which may have to
537.4.1.2 be disconnected rapidly from the supply to prevent or remove danger. Where there is a risk of electric shock the emergency switch is to disconnect all live conductors, except in three-phase TN-S and TN-C-S systems where the neutral need not be switched.

537.4.1.3 The means of emergency switching must act as directly as possible on the appropriate supply conductors and the arrangement must be such that one single action only will interrupt the appropriate supply.

537.4.2.8 A plug and socket-outlet or similar device must not be selected as a device for emergency switching.

An emergency switch must be:

537.4.2.1   i   capable of cutting off the full load current, taking account of stalled motor currents where appropriate

ii hand operated and directly interrupt the main circuit where practicable 537.4.2.3
iii clearly identified, preferably by colour. If a colour is used, this should be red with a contrasting background 537.4.2.4
iv readily accessible at the place where danger may occur and, where appropriate, at any additional remote position from which that danger can be removed 537.4.2.5
v of the latching type or capable of being restrained in the 'OFF' or 'STOP' position, unless both the means of operation and re-energizing are under the control of the same person. The release of an emergency switching device must not re-energize the relevant part of the installation; it must be necessary to take a further action, such as pushing a 'start' button 537.4.2.6
vi so placed and durably marked so as to be readily identifiable and convenient for its intended use. 537.4.2.7

5.4 Functional switching 537.5

A switch must be installed in each part of a circuit which may require to be controlled independently of other parts of the installation. 537.5.1.1

Switches must not be installed in the neutral conductor alone. 537.5.1.2

All current-using equipment requiring control shall be supplied via a switch. 537.5.1.3

Off-load isolators, fuses and links must not be used for functional switching. 537.5.2.3

5.5 Firefighter's switches 537.6

A firefighter's switch must be provided to disconnect the supply to any external electrical installation operating at a voltage exceeding low voltage, for example a neon sign or any interior discharge lighting installation operating at a voltage exceeding low voltage. 537.6.1

Such installations are outside the scope of this Guide (see Regulations 537.6.1 to 537.6.4 of BS 7671:2008).

Labelling 6

6.1 Labels to be provided

The following durable labels are to be securely fixed on or adjacent to installed equipment.

i Unexpected presence of nominal voltage (U_0) exceeding 230 V   514.10.1

Where the nominal voltage (U_0) exceeds 230 V, and it would not normally be expected to be so high, a warning label stating the maximum voltage present must be provided where it can be seen before gaining access to live parts.

ii Nominal voltage exceeding 230 volts (U_0) between simultaneously accessible equipment   514.10.1

For simultaneously accessible equipment with terminals or other fixed live parts having a nominal voltage (U_0) exceeding 230 volts between them, a warning label must be provided where it can be seen before gaining access to live parts.

iii Presence of different nominal voltages in the same equipment   514.10.1

Where equipment contains different nominal voltages, e.g. both low and extra-low, a warning label stating the voltages present must be provided so that it can be seen before gaining access to simultaneously accessible live parts.

iv Connection of earthing and bonding conductors   514.13.1

▼ **Figure 6.1**   Label at connection of earthing and bonding conductors

On-Site Guide | **37**
© The Institution of Engineering and Technology

A permanent label to BS 951 (Figure 6.1) must be permanently fixed in a visible position at or near the point of connection of

1 every earthing conductor to an earth electrode,
2 every protective bonding conductor to extraneous-conductive-parts, and
3 at the main earth terminal, where it is not part of the main switchgear.

514.1.1 **v Purpose of switchgear and controlgear**

Unless there is no possibility of confusion, a label indicating the purpose of each item of switchgear and controlgear must be fixed on or adjacent to the gear. It may be necessary to label the item controlled, as well as its controlgear.

514.8.1 **vi Identification of protective devices**

A protective device, e.g. fuse or circuit-breaker, must be arranged and identified so that the circuit protected may be easily recognised.

537.2.2.6 **vii Identification of isolators**
514.11.1

Where it is not immediately apparent, all isolating devices must be clearly identified by position or durable marking. The location of each disconnector or isolator must be indicated unless there is no possibility of confusion.

514.11.1 **viii Isolation requiring more than one device**

A durable warning notice must be permanently fixed in a clearly visible position to identify the appropriate isolating devices, where equipment or an enclosure contains live parts which cannot be isolated by a single device.

514.12.1 **ix Periodic inspection and testing**

A notice of durable material indelibly marked with the words as Figure 6.2 must be fixed in a prominent position at or near the origin of every installation. The person carrying out the initial verification must complete the notice, and it must be updated after each periodic inspection.

▼ **Figure 6.2** Label for periodic inspection and testing

IMPORTANT

This installation should be periodically inspected and tested and a report on its condition obtained, as prescribed in the IEE Wiring Regulations BS 7671 Requirements for Electrical Installations.

Date of last inspection ....

Recommended date of next inspection ....

6

x Diagrams   514.9

A diagram, chart or schedule must be provided indicating:

a the number of points, size and type of cables for each circuit,
b the method of providing protection against electric shock, and
c any circuit vulnerable to an insulation test.

The schedules of test results (Form 4) of Appendix 7 meet the above requirement for a schedule.

For simple installations the foregoing information may be given in a schedule, with a durable copy provided within or adjacent to each distribution board or consumer unit.

xi Residual current devices   514.12.2

Where an installation incorporates an RCD, a notice with the words in Figure 6.3 (and no smaller than the example shown in BS 7671:2008) must be fixed in a permanent position at or near the origin of the installation.

▼ **Figure 6.3**   Label for the testing of a residual current device

> This installation, or part of it, is protected by a device which automatically switches off the power supply if an earth fault develops. **Test quarterly** by pressing the button marked **'T'** or **'Test'.** The device should switch off the supply and should be then switched on to restore the supply. If the device does not switch off the supply when the button is pressed seek expert advice.

xii Warning notice – non-standard colours   514.14.1

If additions or alterations are made to an installation so that some of the wiring complies with the harmonized colours of Table 11A in Appendix 11 and there is also wiring in the earlier colours, a warning notice must be affixed at or near the appropriate distribution board with the wording in Figure 6.4.

▼ **Figure 6.4** Label advising of wiring colours to two versions of BS 7671

> **CAUTION**
> This installation has wiring colours to two versions of BS 7671.
> Great care should be taken before undertaking extension, alteration or repair that all conductors are correctly identified.

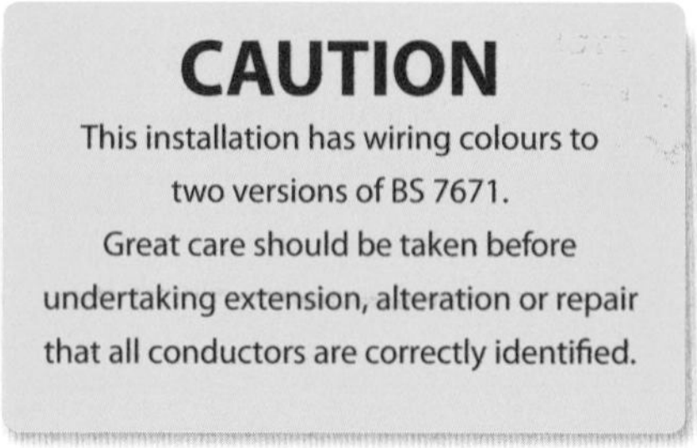

514.15.1 **xiii Warning notice – dual supply**

Where an installation includes a generating set, such as a small-scale embedded generator (SSEG), which is used as an additional source of supply in parallel with another source, normally the distributor's supply, warning notices must be affixed at the following locations in the installation:

- **a** at the origin of the installation
- **b** at the meter position, if remote from the meter
- **c** at the consumer unit or distribution board to which the generating set is connected
- **d** at all points of isolation of both sources of supply.

The warning notice must have the wording in Figure 6.5.

▼ **Figure 6.5** Label advising of dual supply

514.16 **xiv Warning notice – high protective conductor current**

543.7.1.5 At the distribution board, information must be provided indicating those circuits having a high protective conductor current. This information must be positioned so as to be visible to a person who is modifying or extending the circuit (Figure 6.6).

▼ **Figure 6.6** Label advising of high protective conductor current

> **WARNING**
> **HIGH PROTECTIVE CONDUCTOR CURRENT**
> The following circuits have a high protective conductor current:
>
> ....
> ....

xv **Warning notice – photovoltaic systems** 712.537.2.2.5.1

All junction boxes (PV generator and PV array boxes) must carry a warning label indicating that parts inside the boxes may still be live after isolation from the PV convertor (Figure 6.7).

▼ **Figure 6.7** Label advising of live parts within enclosures in a PV system

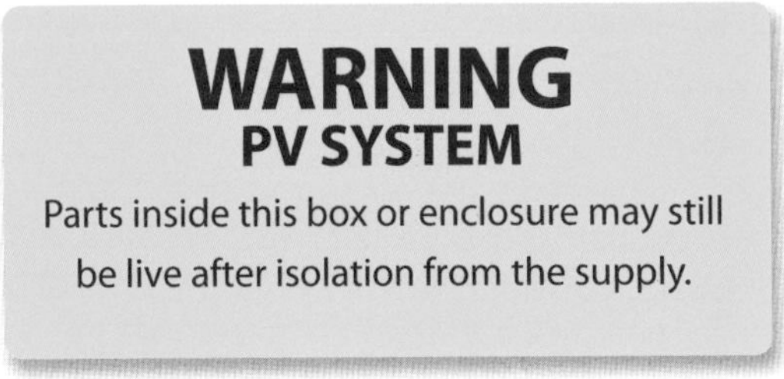

> **WARNING**
> **PV SYSTEM**
> Parts inside this box or enclosure may still be live after isolation from the supply.

Final circuits 7

7.1 Final circuits

Table 7.1 has been designed to enable a radial or ring final circuit to be installed without calculation where the supply is at 230 V single-phase or 400 V three-phase. For other voltages, the maximum circuit length given in the table must be corrected by the application of the formula

411.3.2
411.3.3
525.3
Appx 12
Appx 4
Table 4D5

$$L_p = \frac{L_t \times U_0}{230}$$

where:

L_p is the permitted length for voltage U_0
L_t is the tabulated length for 230 V
U_0 is the supply voltage.

The conditions assumed are that:

i the installation is supplied by

    a a TN-C-S system with a maximum external earth fault loop impedance, Z_e, of 0.35 Ω, or

    b a TN-S system with a maximum Z_e of 0.8 Ω, or

    c a TT system with RCDs installed as described in 3.6

ii the final circuit is connected to a distribution board or consumer unit at the origin of the installation

iii the installation method is listed in column 4 of Table 7.1

iv the ambient temperature throughout the length of the circuit does not exceed 30 °C

v the characteristics of protective devices are in accordance with Appendix 3 of BS 7671

vi the cable conductors are of copper

vii for other than lighting circuits, the voltage drop must not exceed 5 per cent

viii a disconnection time of 0.4 s is applicable for all circuits up to and including 32 A rating and 5 s for all others.

Table 4C1

Appx 3

Appx 12

▼ **Table 7.1** Maximum cable length for a 230 V final circuit in domestic premises and similar using 70 °C thermoplastic (PVC) insulated and sheathed flat cable

Protective device		Cable size (mm²)	Allowed installation methods (note 2)	Maximum length (m) (note 1)			
				$Z_s \leq 0.8\,\Omega$ TN-S		$Z_s \leq 0.35\,\Omega$ TN-C-S	
Rating (A)	Type			RCD (note 4)	No RCD	RCD (note 4)	No RCD
1	2	3	4	5	6	7	8
Ring circuits (5% voltage drop, load distributed)							
30	BS 1361	2.5/1.5	A,100,102,C	111	59zs	111	111
	BS 3036		A,100,102,C	111	49zs	111	111
32	BS 88-2.2, BS 88-6	2.5/1.5	A,100,102,C	106	41zs	106	106
	cb/RCBO Type B			106	106	106	106
	cb/RCBO Type C		A,100,102,C	NPsc	NPzs	82sc	63zs
	cb/RCBO Type D			NPsc	NPzs	2sc	1zs
30	BS 1361	4.0/1.5	101,A,100,102,C	183	69zs	183	159zs
	BS 3036		101,A,100,102,C	183	57zs	183	147zs
32	BS 88-2.2, BS 88-6	4.0/1.5	101,A,100,102,C	176	47zs	176	137zs
	cb/RCBO Type B			176	127zs	176	73zs
	cb/RCBO Type C		101,A,100,102,C	NPsc	NPzs	133sc	73zs
	cb/RCBO Type D			NPsc	NPzs	3sc	1zs
Lighting circuits (3% voltage drop, load distributed)							
5	BS 1361	1.0/1.0	103,101,A,100,102,C	71	71	71	71
	BS 3036		103,101,A,100,102,C	71	71	71	71
5	BS 1361	1.5/1.0	103,101,A,100,102,C	108	108	108	108
	BS 3036		103,101,A,100,102,C	108	108	108	108
6	BS 88-2.2, BS 88-6	1.0/1.0	103,101,A,100,102,C	59	59	59	59
	cb/RCBO Type B			59	59	59	59
	cb/RCBO Type C		103,101,A,100,102,C	59	59	59	59
	cb/RCBO Type D			25sc	25zs	36sc	36zs

▼ **Table 7.1** *continued*

Protective device		Cable size (mm²)	Allowed installation methods (note 2)	Maximum length (m) (note 1)			
				$Z_s \leq 0.8\,\Omega$ TN-S		$Z_s \leq 0.35\,\Omega$ TN-C-S	
Rating (A)	Type			RCD (note 4)	No RCD	RCD (note 4)	No RCD
1	2	3	4	5	6	7	8
Lighting circuits (3% voltage drop, load distributed)							
6	BS 88-2.2, BS 88-6	1.5/1.0	103,101,A,100,102,C	90	90	90	90
	cb/RCBO Type B			90	90	90	90
	cb/RCBO Type C		103,101,A,100,102,C	90	83zs	90	90
	cb/RCBO Type D			38sc	30zs	53sc	43zs
10	BS 88-2.2, BS 88-6	1.0/1.0	101,A,100,102,C	35	35	35	35
	cb/RCBO Type B			35	35	35	35
	cb/RCBO Type C		101,A,100,102,C	34sc	34zs	35	35
	cb/RCBO Type D			8sc	8zs	18sc	18zs
10	BS 88-2.2, BS 88-6	1.5/1.0	103,101,A,100,102,C	52	52	52	52
	cb/RCBO Type B			52	52	52	52
	cb/RCBO Type C		103,101,A,100,102,C	51sc	41zs	52	52
	cb/RCBO Type D			12sc	9zs	27sc	22zs
16	BS 88-2.2, BS 88-6	1.5/1.0	100,102,C	33	33	33	33
	cb/RCBO Type B			33	33	33	33
	cb/RCBO Type C		100,102,C	21sc	17zs	33	30zs
	cb/RCBO Type D			NPsc	NPzs	12sc	10zs
16	BS 88-2.2, BS 88-6	2.5/1.5	101,A,100,102,C	53	53	53	53
	cb/RCBO Type B			53	53	53	53
	cb/RCBO Type C		101,A,100,102,C	35sc	27zs	53	46zs
	cb/RCBO Type D			NPsc	NPzs	20sc	15zs

continues

Table 7.1 continued

Protective device		Cable size (mm²)	Allowed installation methods (note 2)	Maximum length (m) (note 1)				
				$Z_s \leq 0.8\ \Omega$ TN-S			$Z_s \leq 0.35\ \Omega$ TN-C-S	
Rating (A)	Type			RCD (note 4)	No RCD	RCD (note 4)	No RCD	
1	2	3	4	5	6	7	8	

Radial circuits (5% voltage drop, terminal load)

Rating (A)	Type	Cable size (mm²)	Allowed installation methods (note 2)	$Z_s \leq 0.8\ \Omega$ TN-S RCD (note 4)	$Z_s \leq 0.8\ \Omega$ TN-S No RCD	$Z_s \leq 0.35\ \Omega$ TN-C-S RCD (note 4)	$Z_s \leq 0.35\ \Omega$ TN-C-S No RCD
5	BS 1361 BS 3036	1.0/1.0	103,101,A,100,102,C 103,101,A,100,102,C	56 56	56 56	56 56	56 56
5	BS 1361 BS 3036	1.5/1.0	103,101,A,100,102,C 103,101,A,100,102,C	88 88	88 88	88 88	88 88
6	BS 88-2.2, BS 88-6 cb/RCBO Type B cb/RCBO Type C cb/RCBO Type D	1.0/1.0	103,101,A,100,102,C 103,101,A,100,102,C	46 46 46 25sc	46 46 46 25zs	46 46 46 36sc	46 46 46 36zs
6	BS 88-2.2, BS 88-6 cb/RCBO Type B cb/RCBO Type C cb/RCBO Type D	1.5/1.0	103,101,A,100,102,C	72 72 72 38sc	72 72 72 30zs	72 72 72 53sc	72 72 72 43zs
10	BS 88-2.2, BS 88-6 cb/RCBO Type B cb/RCBO Type C cb/RCBO Type D	1.0/1.0	101,A,100,102,C 101,A,100,102,C	26 26 26 8sc	26 26 26 8zs	26 26 26 18sc	26 26 26 18zs
10	BS 88-2.2, BS 88-6 cb/RCBO Type B cb/RCBO Type C cb/RCBO Type D	1.5/1.0	103,101,A,100,102,C 103,101,A,100,102,C	39 39 39 12sc	39 39 39 9zs	39 39 39 27sc	39 39 39 22zs
15	BS 1361 BS 3036	1.0/1.0	C NP	17 NPol	17 NPol	17 NPol	17 NPol

▼ **Table 7.1** *continued*

	Protective device		Cable size (mm²)	Allowed installation methods (note 2)	Maximum length (m) (note 1)			
		Type			$Z_s \leq 0.8\,\Omega$ TN-S		$Z_s \leq 0.35\,\Omega$ TN-C-S	
Rating (A)					RCD (note 4)	No RCD	RCD (note 4)	No RCD
1		2	3	4	5	6	7	8
15	BS 1361 BS 3036		1.5/1.0	100,102,C NP	26 NPol	26 NPol	26 NPol	26 NPol
15	BS 1361 BS 3036		2.5/1.5	101,A,100,102,C 100,102,C	43 45	43 45	43 45	43 45
15	BS 1361 BS 3036		4.0/1.5	103,101,A,100,102,C 101,A,100,102,C	72 75	72 75	72 75	72 75
16	BS 88-2.2, BS 88-6 cb/RCBO Type B cb/RCBO Type C cb/RCBO Type D		1.0/1.0	C C	16 16 14sc NPsc	16 16 14zs NPzs	16 16 8sc	16 16 16 8zs
16	BS 88-2.2, BS 88-6 cb/RCBO Type B cb/RCBO Type C cb/RCBO Type D		1.5/1.0	100,102,C 100,102,C	24 24 21sc NPsc	24 24 17zs NPzs	24 24 24 12sc	24 24 24 10zs
16	BS 88-2.2, BS 88-6 cb/RCBO Type B cb/RCBO Type C cb/RCBO Type D		2.5/1.5	101,A,100,102,C 101,A,100,102,C	40 40 35sc NPsc	40 40 27zs NPzs	40 40 40 20sc	40 40 40 15zs
16	BS 88-2.2, BS 88-6 cb/RCBO Type B cb/RCBO Type C cb/RCBO Type D		4.0/1.5	103,101,A,100,102,C 103,101,A,100,102,C	66 66 57sc NPsc	66 66 31zs NPzs	66 66 66 33sc	66 66 54zs 18zs

continues

▼ Table 7.1 continued

Protective device		Cable size (mm²)	Allowed installation methods (note 2)	Maximum length (m) (note 1)					
				$Z_s \leq 0.8\ \Omega$ TN-S			$Z_s \leq 0.35\ \Omega$ TN-C-S		
Rating (A)	Type			RCD (note 4)	No RCD	RCD (note 4)	No RCD		
1	2	3	4	5	6	7	8		
20	BS 88-2.2, BS 88-6	2.5/1.5	A,100,102,C	31	31	31	31		
	BS 1361		A,100,102,C	31	31	31	31		
	BS 3036		NP	NPol	NPol	NPol	NPol		
	cb/RCBO Type B		A,100,102,C	31	31	31	31		
	cb/RCBO Type C			19sc	14zs	12sc	9zs		
	cb/RCBO Type D			NPsc	NPzs				
20	BS 88-2.2, BS 88-6	4.0/1.5	101,A,100,102,C	53	48zs	53	53		
	BS 1361		101,A,100,102,C	53	44zs	53	53		
	BS 3036		C	57	48zs	57	57		
	cb/RCBO Type B		101,A,100,102,C	53	53	53	53		
	cb/RCBO Type C			31sc	17zs	20sc	39zs		
	cb/RCBO Type D			NPsc	NPzs		11zs		
20	BS 88-2.2, BS 88-6	6.0/2.5	103,101,A,100,102,C	81	77zs	81	81		
	BS 1361		103,101,A,100,102,C	81	71zs	81	81		
	BS 3036		A,100,102,C	85	77zs	85	85		
	cb/RCBO Type B		103,101,A,100,102,C	81	81	81	81		
	cb/RCBO Type C			47sc	27zs	30sc	63zs		
	cb/RCBO Type D			NPsc	NPzs		17zs		
25	BS 88-2.2, BS 88-6	2.5/1.5	C	26	26	26	26		
	BS 1361		C	26	26	26	26		
	BS 3036			6sc	5zs	26	24zs		
	cb/RCBO Type B			NPsc	NPzs	6sc	4zs		
	cb/RCBO Type C								
	cb/RCBO Type D								
25	BS 88-2.2, BS 88-6	4.0/1.5	A,100,102,C	42	31zs	42	42		
	BS 1361		A,100,102,C	42	42	42	42		
	BS 3036			10sc	6zs	42	28zs		
	cb/RCBO Type B			NPsc	NPzs	9sc	5zs		
	cb/RCBO Type C								
	cb/RCBO Type D								

▼ **Table 7.1** *continued*

Rating (A)	Protective device Type	Cable size (mm²)	Allowed installation methods (note 2)	Maximum length (m) (note 1) $Z_s \leq 0.8\,\Omega$ TN-S RCD (note 4)	$Z_s \leq 0.8\,\Omega$ TN-S No RCD	$Z_s \leq 0.35\,\Omega$ TN-C-S RCD (note 4)	$Z_s \leq 0.35\,\Omega$ TN-C-S No RCD
1	2	3	4	5	6	7	8
25	BS 88-2.2, BS 88-6	6.0/2.5	101,A,100,102,C	64	50zs	64	64
	cb/RCBO Type B			64	64	64	64
	cb/RCBO Type C		101,A,100,102,C	16sc	9zs	64	45zs
	cb/RCBO Type D			NPsc	NPzs	14sc	8zs
30	BS 1361	4.0/1.5	C	36	17zs	36	36
	BS 3036		NP	NPol	NPol	NPol	NPol
30	BS 1361	6.0/2.5	A,100,102,C	53	27zs	53	53
	BS 3036		C	57	23zs	57	57
30	BS 1361	10.0/4.0	103,101,A,100,102,C	90	45zs	90	90
	BS 3036		A,100,102,C	93	37zs	93	93
32	BS 88-2.2, BS 88-6	4.0/1.5	C	33	11zs	33	33
	cb/RCBO Type B			33	31zs	33	33
	cb/RCBO Type C		C	NPsc	NPzs	33	18zs
	cb/RCBO Type D			NPsc	NPzs	1sc	NPzs
32	BS 88-2.2, BS 88-6	6.0/2.5	A,100,102,C	49	19zs	49	49
	cb/RCBO Type B			49	49	49	49
	cb/RCBO Type C		A,100,102,C	NPsc	NPzs	49	29zs
	cb/RCBO Type D			NPsc	NPzs	1sc	1zs
32	BS 88-2.2, BS 88-6	10.0/4.0	103,101,A,100,102,C	81	31zs	81	81
	cb/RCBO Type B			81	81	81	81
	cb/RCBO Type C		103,101,A,100,102,C	NPsc	NPzs	81	47zs
	cb/RCBO Type D			NPsc	NPzs	2sc	1zs

continues

▼ **Table 7.1** *continued*

Protective device		Cable size (mm²)	Allowed installation methods (note 2)	Maximum length (m) (note 1)					
				$Z_s \leq 0.8\,\Omega$ TN-S		No RCD	$Z_s \leq 0.35\,\Omega$ TN-C-S		No RCD
Rating (A)	Type			RCD (note 4)			RCD (note 4)		
1	2	3	4	5		6	7		8
40	BS 88-2.2, BS 88-6	6.0/2.5	C	40		27ad	40		40
	cb/RCBO Type B		⎱ C	40		27zs	40		40
	cb/RCBO Type C		⎰	NPsc		NPzs	30sc		17zs
	cb/RCBO Type D			NPsc		NPzs	NPsc		NPzs
40	BS 88-2.2, BS 88-6	10.0/4.0	A,100,102,C	66		66	66		66
	cb/RCBO Type B		⎱	66		45zs	66		66
	cb/RCBO Type C		⎰ A,100,102,C	NPsc		NPzs	51sc		29zs
	cb/RCBO Type D			NPsc		NPzs	NPsc		NPzs
40	BS 88-2.2, BS 88-6	16.0/6.0	103,101,A,100,102,C	104		104	104		104
	cb/RCBO Type B		⎱	104		68zs	104		104
	cb/RCBO Type C		⎰ 103,101,A,100,102,C	NPsc		NPzs	81sc		44zs
	cb/RCBO Type D			NPsc		NPzs	NPsc		NPzs
45	BS 1361	6.0/2.5	C	21sc		NPad	35		22ad
	BS 3036		NP	NPol		NPol	NPol		NPol
45	BS 1361	10.0/4.0	100,102,C	36sc		5ad	58		58
	BS 3036		C	62		62	62		62
45	BS 1361	16.0/6.0	101,A,100,102,C	58sc		31zs	91		91
	BS 3036		102,C	97		97	97		97

Notes to Table 7.1:
1 Voltage drop is the limiting constraint on the circuit cable length unless marked as follows:
   - **ad** Limited by reduced csa of protective conductor (adiabatic limit)
   - **ol** Cable/device/load combination not allowed in any of the installation conditions
   - **zs** Limited by earth fault loop impedance Z_s
   - **sc** Limited by line to neutral loop impedance (short-circuit).
2 The allowed installation methods are listed, see Tables 7.2 and 7.3 for further description.
3 NP Not Permitted, prohibiting factor as note 1.
4 For application of RCDs and RCBOs, see 3.6.2.

▼ **Table 7.2** Installation reference methods and cable ratings for 70 °C thermoplastic (PVC) insulated and sheathed flat cable with protective conductor

Installation reference method		Conductor cross-sectional area (mm²)						
Ref.	Description	1.0	1.5	2.5	4	6	10	16
		A	A	A	A	A	A	A
C	Clipped direct	16	20	27	37	47	64	85
B*	Enclosed in conduit or trunking on a wall, etc.	13	16.5	23	30	38	52	69
102	In a stud wall with thermal insulation with cable touching the wall	13	16	21	27	35	47	63
100	In contact with plasterboard ceiling or joists covered by thermal insulation not exceeding 100 mm	13	16	21	27	34	45	57
A	Enclosed in conduit in an insulated wall	11.5	14.5	20	26	32	44	57
101	In contact with plasterboard ceiling or joists covered by thermal insulation exceeding 100 mm	10.5	13	17	22	27	36	46
103	Surrounded by thermal insulation including in a stud wall with thermal insulation with cable not touching a wall	8	10	13.5	17.5	23.5	32	42.5

Notes:
1 Cable ratings taken from Table 4D5 of BS 7671.
2 B* taken from Table 4D2A of BS 7671, see Appendix 6.

▼ **Table 7.3** Installation methods specifically for flat twin and earth cables in thermal insulation

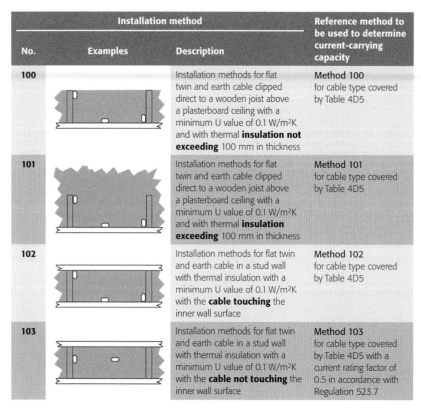

No.	Examples	Description	Reference method to be used to determine current-carrying capacity
100		Installation methods for flat twin and earth cable clipped direct to a wooden joist above a plasterboard ceiling with a minimum U value of 0.1 W/m²K and with thermal **insulation not exceeding** 100 mm in thickness	Method 100 for cable type covered by Table 4D5
101		Installation methods for flat twin and earth cable clipped direct to a wooden joist above a plasterboard ceiling with a minimum U value of 0.1 W/m²K and with thermal **insulation exceeding** 100 mm in thickness	Method 101 for cable type covered by Table 4D5
102		Installation methods for flat twin and earth cable in a stud wall with thermal insulation with a minimum U value of 0.1 W/m²K with the **cable touching** the inner wall surface	Method 102 for cable type covered by Table 4D5
103		Installation methods for flat twin and earth cable in a stud wall with thermal insulation with a minimum U value of 0.1 W/m²K with the **cable not touching** the inner wall surface	Method 103 for cable type covered by Table 4D5 with a current rating factor of 0.5 in accordance with Regulation 523.7

Notes:
1. Wherever practicable, a cable should be fixed in a position such that it will not be covered with thermal insulation.
2. Regulation 523.7, BS 5803-5: Appendix C 'Avoidance of overheating of electric cables', Building Regulations Approved Document B, and Thermal Insulation: avoiding risks, BR 262, BRE 2001 refer.

7.2 Standard final circuits

7.2.1 Grouping of circuit cables

The tables assume heating (including water heating) cables are not grouped.

For cables of household or similar installations (heating and water heating excepted), if the following rules are followed derating for grouping is not necessary:

    i Cables are not grouped, that is, they are separated by at least two cable diameters when installed under insulation, namely installation methods 100, 101, 102 and 103.
    ii Cables clipped direct (including in cement or plaster) are clipped side by side in one layer separated by at least one cable diameter.
    iii Cables above ceilings are clipped to joists as per installation reference methods 100 to 103 of Table 2A2 of BS 7671.

For other groupings, ambient temperatures higher than 30 °C or enclosure in thermal insulation, cable csa will need to be increased as per Appendix 6 of this Guide.

7.2.2 Socket-outlet circuits

The length represents the total ring cable loop length and does not include any spurs.

As a rule of thumb for rings, unfused spur lengths should not exceed 1/8 the cable length from the spur to the furthest point of the ring.

The total number of fused spurs is unlimited but the number of non-fused spurs is not to exceed the total number of socket-outlets and items of stationary equipment connected directly in the circuit.

A non-fused spur feeds only one twin or multiple socket-outlet or one permanently connected item of electrical equipment. Such a spur is connected to a circuit at the terminals of socket-outlets or at junction boxes or at the origin of the circuit in the distribution board.

A fused spur is connected to the circuit through a fused connection unit, the rating of the fuse in the unit not exceeding that of the cable forming the spur and, in any event, not exceeding 13 A. The number of socket-outlets which may be supplied by a fused spur is unlimited.

The circuit is assumed to have a load of 20 A at the furthest point and the balance to the rating of the protective device evenly distributed. (For a 32 A device this equates to a load of 26 A at the furthest point.)

7.2.3 Lighting circuits

A maximum voltage drop of 3 per cent of the 230 V nominal supply voltage has been allowed in the circuits (see Appendix 6).

The circuit is assumed to have a load equal to the rated current (I_n) of the circuit protective device, evenly distributed along the circuit. Where this is not the case, circuit lengths will need to be reduced where voltage drop is the limiting factor, or halved where load is all at the extremity.

The most onerous installation condition acceptable for the load and device rating is presumed when calculating the limiting voltage drop. If the installation conditions are not the most onerous allowed (see column 4 of Table 7.1) the voltage drop will not be as great as presumed in the table.

7.2.4 RCDs

Where circuits have residual current protection, the limiting factor is often the maximum loop impedance that will result in operation of the overcurrent device within 5 seconds for a short-circuit (line to neutral) fault. (See note 1 to Table 7.1 and limiting factor sc.)

7.2.5 Requirement for RCDs

RCDs are required:

i	where the earth fault loop impedance is too high to provide the required disconnection time, e.g. where the distributor does not provide an earth – TT systems	411.5
ii	for socket-outlet circuits in domestic and similar installations	411.3.3
iii	for circuits of locations containing a bath or shower	701.411.3.3
iv	for circuits supplying mobile equipment not exceeding 32 A for use outdoors	411.3.3
v	for cables without earthed metallic covering installed in walls or partitions at a depth of less than 50 mm and not protected by earthed steel conduit or similar	522.6.7
vi	for cables without earthed metallic covering installed in walls or partitions with metal parts (excluding screws or nails) and not protected by earthed steel conduit or the like.	522.6.8

30 mA RCDs are required for **ii** to **vi** above.

411.3.3 RCDs may be omitted for:

a specific labelled sockets, such as a socket for a freezer. However, the circuit cables must not require RCDs as per **v** and **vi** above, that is, circuit cables must be enclosed in earthed steel conduit or have an earthed metal sheath or be at a depth of 50 mm in a wall or partition without metal parts

b socket-outlet circuits in industrial and commercial premises where the use of equipment and work on the building fabric and electrical installation is under the supervision of skilled or instructed persons.

See 3.6.

7.2.6 TT systems

For TT systems the figures for TN-C-S systems, with RCDs, may be used provided that:

i the circuit is protected by an RCD to BS 4293, BS EN 61008 or BS EN 61009 with a rated residual operating current not exceeding that required for its circuit position,
ii the total earth fault loop impedance is verified as being less than 200 Ω, and
iii a device giving both overload and short-circuit protection is installed in the circuit. This may be an RCBO or a combination of a fuse or circuit-breaker with an RCD.

7.2.7 Choice of protective device

The selection of protective device depends upon:

i prospective fault current
ii circuit load characteristics
iii cable current-carrying capacity
iv disconnection time limit.

While these factors have generally been allowed for in the standard final circuits in Table 7.1, the following additional guidance is given:

i Prospective fault current

434.5.1 If a protective device is to operate safely, its rated short-circuit capacity must not be less than the prospective fault current at the point where it is installed. See Table 7.4.

313.1 The distributor needs to be consulted as to the prospective fault current at the origin of the installation. Except for London and some other major city centres, the maximum fault current for 230 V single-phase supplies up to 100 A will not exceed 16 kA. In general, the fault current is unlikely to exceed 16.5 kA.

Consumer units incorporating protective devices complying as a whole assembly with BS 5486-13 or BS EN 60439-3 are suitable for locations with fault currents up to 16 kA when supplied through a type II fuse to BS 1361:1971 (1992) or BS 88 fuse rated at no more than 100 A.

▼ **Table 7.4** Rated short-circuit capacities

Device type	Device designation	Rated short-circuit capacity (kA)	
Semi-enclosed fuse to BS 3036 with category of duty	S1A S2A S4A	1 2 4	
Cartridge fuse to BS 1361 type I type II		16.5 33.0	
General purpose fuse to BS 88-2		50 at 415 V	
General purpose fuse to BS 88-6		16.5 at 240 V 80 at 415 V	
Circuit-breakers to BS 3871 (replaced by BS EN 60898)	M1 M1.5 M3 M4.5 M6 M9	1 1.5 3 4.5 6 9	
Circuit-breakers to BS EN 60898* and RCBOs to BS EN 61009		I_{cn} 1.5 3.0 6 10 15 20 25	I_{cs} (1.5) (3.0) (6.0) (7.5) (7.5) (10.0) (12.5)

* Two short-circuit capacities are defined in BS EN 60898 and BS EN 61009:

I_{cn} the rated short-circuit capacity (marked on the device).
I_{cs} the in-service short-circuit capacity.

The difference between the two is the condition of the circuit-breaker after manufacturer's testing.

I_{cn} is the maximum fault current the breaker can interrupt safely, although the breaker may no longer be usable.

I_{cs} is the maximum fault current the breaker can interrupt safely without loss of performance.

The I_{cn} value (in amperes) is normally marked on the device in a rectangle, e.g. 6000 and for the majority of applications the prospective fault current at the terminals of the circuit-breaker should not exceed this value.

For domestic installations the prospective fault current is unlikely to exceed 6 kA, up to which value the I_{cn} will equal I_{cs}.

The short-circuit capacity of devices to BS EN 60947-2 is as specified by the manufacturer.

ii Circuit load characteristics

533.1.1.3
a *Semi-enclosed fuses.* Fuses should preferably be of the cartridge type. However, semi-enclosed fuses to BS 3036 are still permitted for use in domestic and similar premises if fitted with a fuse element which, in the absence of more specific advice from the manufacturer, meets the requirements of Table 53.1.
b *Cartridge fuses to BS 1361.* These are for use in domestic and similar premises.
c *Cartridge fuses to BS 88.* Three types are specified:

> gG fuse links with a full-range breaking capacity for general application
> gM fuse links with a full-range breaking capacity for the protection of motor circuits
> aM fuse links for the protection of motor circuits

d Circuit-breakers to BS EN 60898 (or BS 3871-1) and RCBOs to BS EN 61009. Guidance on selection is given in Table 7.5.

▼ **Table 7.5** Application of circuit-breakers

Circuit-breaker type	Trip current (0.1 s to 5 s)	Application
1 B	2.7 to 4 I_n 3 to 5 I_n	Domestic and commercial installations having little or no switching surge
2 C 3	4 to 7 I_n 5 to 10 I_n 7 to 10 I_n	General use in commercial/industrial installations where the use of fluorescent lighting, small motors, etc., can produce switching surges that would operate a Type 1 or B circuit-breaker. Type C or 3 may be necessary in highly inductive circuits such as banks of fluorescent lighting
4 D	10 to 50 I_n 10 to 20 I_n	Not suitable for general use Suitable for transformers, X-ray machines, industrial welding equipment, etc., where high inrush currents may occur

Note: I_n is the nominal rating of the circuit-breaker.

Appx 3
Appx 4

iii Cable current-carrying capacities

For guidance on the coordination of device and cable ratings see Appendix 6.

iv Disconnection times

411.3.2.2
411.3.2.3
411.3.2.4
The protective device must operate within 0.2, 0.4, 1 or 5 seconds as appropriate for the circuit. Appendix 2 provides maximum permissible measured earth fault loop impedances for fuses, circuit-breakers and RCBOs.

7.3 Installation considerations

7.3.1 Floors and ceilings

522.6.5

Where a cable is installed under a floor or above a ceiling it shall be run in such a position that it is not liable to be damaged by contact with the floor or ceiling or the fixings thereof.

A cable passing through a joist or ceiling support shall:

   i   be at least 50 mm from the top or bottom, as appropriate, or
   ii have earthed armouring or an earthed metal sheath, or
   iii be enclosed in earthed steel conduit or trunking, or
   iv be provided with mechanical protection sufficient to prevent penetration of the cable by nails, screws and the like (Note: the requirement to prevent penetration is difficult to meet).

See Figure 7.1.

▼ **Figure 7.1** Cables through joists

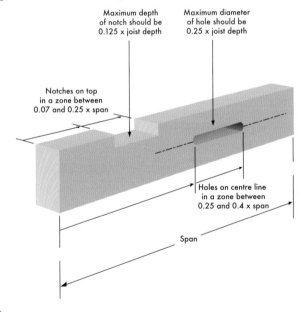

Notes:
1 Maximum diameter of hole should be 0.25 x joist depth.
2 Holes on centre line in a zone between 0.25 and 0.4 x span.
3 Maximum depth of notch should be 0.125 x joist depth.
4 Notches on top in a zone between 0.07 and 0.25 x span.
5 Holes in the same joist should be at least 3 diameters apart.

7.3.2 Walls and partitions

522.6.6

A cable concealed in a wall or partition must:

i be at least 50 mm from the surface, or
ii have earthed armouring or an earthed metal sheath, or
iii be enclosed in earthed steel conduit or trunking, or
iv be provided with mechanical protection sufficient to prevent penetration of the cable by nails, screws and the like (Note: the requirement to prevent penetration is difficult to meet), or
v be installed either horizontally within 150 mm of the top of the wall or partition or vertically within 150 mm of the angle formed by two walls, or run horizontally or vertically to an accessory or consumer unit (see Figure 7.2).

In domestic and similar installations, cables not installed as per **i, ii, iii** or **iv** but complying with **v** shall be protected by a 30 mA RCD.

In domestic and similar installations, cables installed in walls or partitions with a metal or part metal construction shall be either:

a installed as **ii, iii** or **iv** above, or
b protected by a 30 mA RCD.

For installations under the supervision of a skilled or instructed person, such as commercial or industrial where only authorized equipment is used and only skilled persons will work on the building, RCD protection as described above is not required.

Note: Domestic or similar are not considered to be under the supervision of skilled or instructed persons.

▼ **Figure 7.2**   Zones prescribed in Regulation 522.6.6(v) (see **v** above)

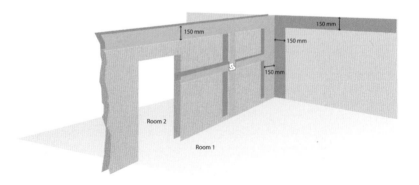

7.4 Proximity to electrical and other services

528

Electrical and all other services must be protected from any harmful mutual effects foreseen as likely under conditions of normal service. For example, cables should not be in contact with or run alongside hot pipes.

528.3

7.4.1 Segregation of Band I and Band II circuits

Band I (extra-low voltage) circuits shall not be contained within the same wiring system (e.g. trunking) as Band II (low voltage) circuits unless:

528.1

    i every cable is insulated for the highest voltage present, or
   ii each conductor of a multicore cable is insulated for the highest voltage present, or
  iii the cables are installed in separate compartments, or
  iv the cables fixed to a cable tray are separated by a partition, or
   v for a multicore cable, they are separated by an earthed metal screen of equivalent current-carrying capacity to that of the largest Band II circuit.

Definitions of voltage bands

Part 2

▶ **Band I circuit:** Circuit that is nominally extra-low voltage, i.e. not exceeding 50 V a.c. or 120 V d.c. For example, SELV, PELV, telecommunications, data and signalling.

▶ **Band II circuit:** Circuit that is nominally low voltage, i.e. 51 to 1000 V a.c. and 121 to 1500 V d.c. Telecommunication cables that are generally ELV but have ringing voltages exceeding 50 V are Band I.

Note: Fire alarm and emergency lighting circuits must be separated from other cables and from each other, in compliance with BS 5839 and BS 5266.

528.1, Note 2

7.4.2 Proximity to communications cables

528.2

An adequate separation between telecommunication wiring (Band I) and electric power and lighting (Band II) circuits must be maintained. This is to prevent mains voltage appearing in telecommunication circuits with consequent danger to personnel. BS 6701:2004 recommends that the minimum separation distances given in Tables 7.6 and 7.7 should be maintained.

▼ **Table 7.6** External cables

Minimum separation distances between external low voltage electricity supply cables operating in excess of 50 V a.c. or 120 V d.c. to earth, but not exceeding 600 V a.c. or 900 V d.c. to earth (Band II), and telecommunications cables (Band I).

Voltage to earth	Normal separation distances	Exceptions to normal separation distances, plus conditions to exception
Exceeding 50 V a.c. or 120 V d.c., but not exceeding 600 V a.c. or 900 V d.c.	50 mm	Below this figure a non-conducting divider should be inserted between the cables

▼ **Table 7.7** Internal cables

Minimum separation distances between internal low voltage electricity supply cables operating in excess of 50 V a.c. or 120 V d.c. to earth, but not exceeding 600 V a.c. or 900 V d.c. to earth (Band II) and telecommunications cables (Band I).

Voltage to earth	Normal separation distances	Exceptions to normal separation distances, plus conditions to exception
Exceeding 50 V a.c. or 120 V d.c., but not exceeding 600 V a.c. or 900 V d.c.	50 mm	50 mm separation need not be maintained, provided that (i) the LV cables are enclosed in separate conduit which, if metallic, is earthed in accordance with BS 7671, **OR** (ii) the LV cables are enclosed in separate trunking which, if metallic, is earthed in accordance with BS 7671, **OR** (iii) the LV cable is of the mineral insulated type or is of earthed armoured construction.

Notes:
1. Where the LV cables share the same tray then the normal separation should be met.
2. Where LV and telecommunications cables are obliged to cross, additional insulation should be provided at the crossing point; this is not necessary if either cable is armoured.

7.4.3 Separation from gas services

(BS 6891:2005 *Installation of low pressure gas pipework in domestic premises*, clause 8.16.2)

Where gas installation pipes are not separated by electrical insulating material from electrical equipment, including cables, they are to be spaced as follows:

a at least 150 mm away from electrical equipment (meters, controls, accessories, distribution boards or consumer units);

b at least 25 mm away from electricity cables.

Further information is given in 2.3.

7.4.4 Induction loops

A particular form of harmful effect may occur when an electrical installation shares the space occupied by a hearing aid induction loop.

Under these circumstances, if line and neutral conductors or switch feeds and switch wires are not run close together, there may be interference with the induction loop.

This can occur when a conventional two-way lighting circuit is installed. This effect can be reduced by connecting as shown in Figure 7.3.

▼ **Figure 7.3**   Circuit for reducing interference with induction loop

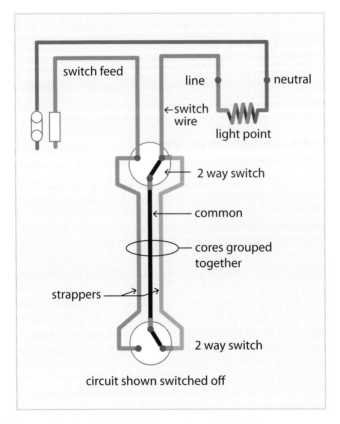

Note: Black/grey switch wires to be identified in accordance with Table 11A.   Table 51

7.5 Compliance with the Building Regulations

In this publication, guidance is limited to accessibility of accessories and brief guidance on smoke and heat alarms in standard houses. Further guidance is given in the IEE publication *Electrician's Guide to the Building Regulations*.

7.5.1 Height of switches, socket-outlets, etc., in dwellings (Part M)

Accessories and controls for general use, such as light switches and socket-outlets, are required by Part M of the Building Regulations to be located so that they can be used by people whose reach is limited. A way of satisfying this requirement is to install socket-outlets and controls throughout the dwelling at a height of between 450 mm and 1200 mm from finished floor level. See Figure 8A in Appendix 8. Because of the sensitivity of circuit-breakers and RCDs fitted to consumer units, consumer units should be readily accessible.

The guidance given in Approved Document M applies to all new dwellings. Note that if a dwelling is rewired there is no requirement to provide the measures described above providing that upon completion the building is no worse in terms of the level of compliance with the other Parts of Schedule 1 to the Building Regulations.

7.5.2 Smoke and heat alarms (Part B)

General

Part B of the Building Regulations and the Building Standards Scotland requires all new and refurbished dwelling houses to be provided with a fire detection and alarm system.

In a standard house (single storey or multi-storey with no storey exceeding 200 m² floor area), the basic requirement can be met by installing interlinked smoke alarms as follows:

1. in circulation areas between sleeping places and places where fires are most likely to start, e.g. kitchens and living rooms
2. in circulation spaces within 7.5 m of the door to each habitable room
3. at least one on every storey
4. if the kitchen is not separated from the circulation area by a door, a compatible interlinked heat detector or heat alarm must additionally be installed in the kitchen.

See Figure 7.4.

▼ **Figure 7.4** Minimum requirement for smoke alarms for a standard house, no storey exceeding 200 m² floor area

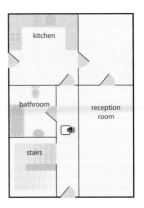
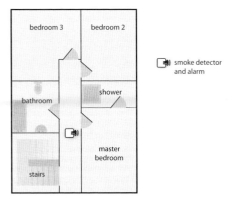

Positioning of equipment

Equipment should be positioned as follows:

 i ceiling mounted alarms and detectors must be fixed at least 300 mm from the walls and luminaires.
 ii the sensors mounted between 25 mm and 600 mm below the ceiling (25 to 150 mm for heat detectors).
 iii only alarms and detectors suitable for wall mounting are to be wall mounted and are to be fixed above doorway height.
 iv smoke alarms should not be fixed next to or directly above heaters or air conditioning outlets. They should not be fixed in bathrooms, showers, cooking areas or garages, or any other place where steam, condensation or fumes could give false alarms.
 v smoke alarms should not be fitted in places that get very hot (such as a boiler room) or very cold (such as an unheated porch).
 vi all equipment should be safely accessible for routine maintenance including testing and cleaning. They should not be fixed over stairs or openings between floors.

Wiring of smoke and heat alarms

The detectors and alarms are required to:

 a be linked so that the operation of one will initiate all units (mains powered smoke detectors may be interlinked by radio)
 b be permanently wired with an independent circuit from the distribution board (consumer unit), or supplied from a local, regularly used lighting circuit (there should be a means of isolating the supply to the alarms without affecting the lighting)

c have a standby power supply, such as a battery or capacitor.

Note: Where all circuits are protected by RCDs there is advantage in supplying fire detectors and alarms from regularly used lighting circuits.

Other than for large houses the cables for the power supply to each self-contained unit and for the interconnections between self-contained units need have no fire survival properties and needs no special segregation.

Otherwise, fire alarm system cables generally are required to be fire resistant and segregated as per BS 5839-1 and BS 5839-6 to minimize adverse effects from:

- Installation cable faults
- Fire on other circuits
- Electromagnetic interference
- Mechanical damage.

Note: Further guidance is given in the IEE publication *Electrician's Guide to the Building Regulations*.

7.6 Earthing requirements for the installation of equipment having high protective conductor current

543.7

7.6.1 Equipment

543.7.1.1 Equipment having a protective conductor current exceeding 3.5 mA but not exceeding 10 mA must be either permanently connected to the fixed wiring of the installation or connected by means of an industrial plug and socket complying with BS EN 60309-2.

543.7.1.2 Equipment having a protective conductor current exceeding 10 mA should be connected by one of the following methods:

  i permanently connected to the wiring of the installation, with the protective conductor selected in accordance with Regulation 543.7.1.3. The permanent connection to the wiring may be by means of a flexible cable
  ii a flexible cable with an industrial plug and socket to BS EN 60309-2, provided that either:

  a the protective conductor of the associated flexible cable is of cross-sectional area not less than 2.5 mm^2 for plugs up to 16 A and not less than 4 mm^2 for plugs rated above 16 A, or
  b the protective conductor of the associated flexible cable is of cross-sectional area not less than that of the line conductor

  iii a protective conductor complying with Section 543 with an earth monitoring system to BS 4444 installed which, in the event of a continuity fault occurring in the protective conductor, automatically disconnects the supply to the equipment.

7.6.2 Circuits

The wiring of every final circuit and distribution circuit having a protective conductor current likely to exceed 10 mA must have high integrity protective conductor connections complying with one or more of the following:

543.7.1.3

   i a single protective conductor having a cross-sectional area not less than 10 mm², complying with Regulations 543.2 and 543.3

   ii a single copper protective conductor having a csa not less than 4 mm², complying with Regulations 543.2 and 543.3, the protective conductor being enclosed to provide additional protection against mechanical damage, for example within a flexible conduit

   iii two individual protective conductors, each complying with Section 543, the ends being terminated independently.

543.7.1.4

   iv earth monitoring or use of double-wound transformer.

543.7.1.3

Note: Distribution boards are to indicate circuits with high protective conductor currents (see 6.1(xiv)).

543.7.1.5

7.6.3 Socket-outlet final circuits

For a final circuit with socket-outlets or connection units, where the protective conductor current in normal service is likely to exceed 10 mA, the following arrangements are acceptable:

543.7.2.1

   i a ring final circuit with a ring protective conductor. Spurs, if provided, require high integrity protective conductor connections (Figure 7.5), or

   ii a radial final circuit with

      a a protective conductor connected as a ring (Figure 7.6), or

      b an additional protective conductor provided by metal conduit or ducting.

▼ **Figure 7.5**   Ring final circuit supplying socket-outlets

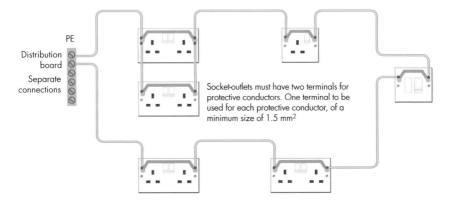

7

▼ **Figure 7.6** Radial final circuit supplying socket-outlets with duplicate protective conductors

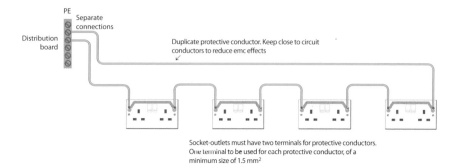

Locations containing a bath or shower   8

8.1 Summary of requirements   701

Because of the presence of water, these locations are onerous for equipment and there is an increased danger of electric shock because of immersion of the body in water.

The additional requirements can be summarised as follows:

i   all the circuits of the location must be protected by 30 mA RCDs   701.411.3.3
ii socket-outlets are not allowed within 3 metres of zone 1 (the edge of the bath   701.512.3
    or shower basin)
iii protection against ingress of water is specified for equipment within the zones,   701.512.2
    see Table 8.1 and Figures 8.1 to 8.3
iv there are restrictions as to where appliances, switchgear and wiring   701.512.3
    accessories may be installed, see Table 8.1 and Figures 8.1 to 8.3.

Supplementary bonding of locations containing a bath or shower is required unless all   701.415.2
the following requirements are met:

▶ all circuits of the location meet the required disconnection times,   411.3.2.2
▶ all circuits of the location are additionally protected by 30 mA RCDs, and   701.411.3.3
▶ all extraneous-conductive parts within the location are effectively connected   411.3.1.2
   via the main protective equipotential bonding to the main earthing terminal.

▼ **Table 8.1** Requirements for equipment (current-using and accessories) in a location containing a bath or shower

701.32
701.5

Zone	Minimum degree of protection	Current-using equipment	Switchgear and accessories
0	IPX7	Only 12 V a.c. rms or 30 V ripple-free d.c. SELV, the safety source installed outside the zones.	None allowed.
1	IPX4 (IPX5 if water jets)	25 V a.c. rms or 60 V ripple-free d.c. SELV or PELV, the safety source installed outside the zones. The following mains voltage fixed, permanently connected equipment allowed: whirlpool units, electric showers, shower pumps, ventilation equipment, towel rails, water heaters, luminaires.	Only 12 V a.c. rms or 30 V ripple-free d.c. SELV switches, the safety source installed outside the zones.
2	IPX4 (IPX5 if water jets)	Fixed permanently connected equipment allowed. General rules apply.	Only switches and sockets of SELV circuits allowed, the source being outside the zones, and shaver supply units complying with BS EN 61558-2-5 if fixed where direct spray is unlikely.
Outside zones	No requirement	General rules apply.	Accessories allowed and SELV socket-outlets and shaver supply units to BS EN 61558-2-5 allowed. Socket-outlets allowed 3 m horizontally from the boundary of zone 1.

601-08-01
601-09-03

601-08-01

8

▼ **Figure 8.1** Zone dimensions in a location containing a bath

Section   Plan

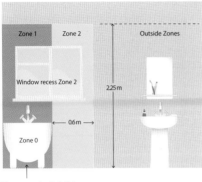 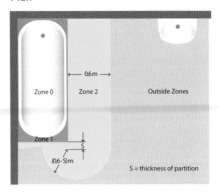

The space under the bath is:
Zone 1 if accessible without the use of a tool
Outside the zones if accessible only with the use of a tool

▼ **Figure 8.2** Zones in a location containing a shower with basin and with permanent fixed partition

Section   Plan

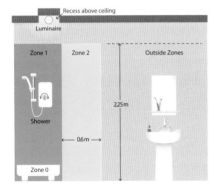 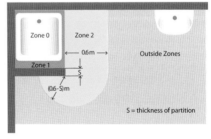

On-Site Guide
© The Institution of Engineering and Technology

▼ **Figure 8.3**   Zones in a location containing a shower without a basin, but with a partition

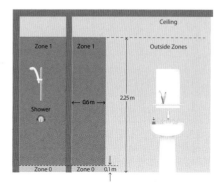
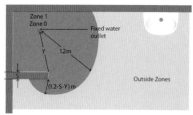

S = thickness of partition
Y = radial distance from the fixed water outlet to the inner corner of the partition

8.2 Underfloor heating

701.753   Underfloor heating installations in these areas should have an overall earthed metallic grid or the heating cable should have an earthed metallic sheath, which must be connected to the protective conductor of the supply circuit.

Note: BS 7671 has further requirements for underfloor heating in Regulation 701.753 and Section 753.

8.3 Shower cubicle in a room used for other purposes

Where a shower cubicle is installed in a room other than a bathroom or shower room the requirements for bathrooms and shower rooms must be complied with.

Inspection and testing 9

9.1 Inspection and testing

Part 6

Every installation must be inspected and tested during erection and on completion before being put into service to verify, so far as is reasonably practicable, that the requirements of the Regulations have been met.

610.1

Precautions must be taken to avoid danger to persons and to avoid damage to property and installed equipment during inspection and testing.

If the inspection and tests are satisfactory, a signed Electrical Installation Certificate together with a Schedule of Inspections and a Schedule of Test Results (as in Appendix 7) are to be given to the person ordering the work.

632.1

9.2 Inspection

611

9.2.1 Procedure and purpose

Inspection must precede testing and must normally be done with that part of the installation under inspection disconnected from the supply.

611.1

The purpose of the inspection is to verify that equipment is:

611.2

- **i** correctly selected and erected in accordance with BS 7671 (and if appropriate its own standard)
- **ii** not visibly damaged or defective so as to impair safety.

9.2.2 Inspection checklist

611.3

The inspection must include at least the checking of relevant items from the following checklist:

- **i** connection of conductors — 526
- **ii** identification of conductors — 514.3
- **iii** routing of cables in safe zones or protection against mechanical damage
- **iv** selection of conductors for current-carrying capacity and voltage drop, in accordance with the design
- **v** connection of single-pole devices for protection or switching in line conductors only — 132.14.1

On-Site Guide | **73**
© The Institution of Engineering and Technology

9

527.2	**vi** correct connection of accessories and equipment (including polarity)
41	**vii** presence of fire barriers, suitable seals and protection against thermal effects
	viii methods of protection against electric shock:

      a basic protection and fault protection, i.e.

414	▶ SELV
	▶ PELV
412	▶ double insulation
	▶ reinforced insulation

      b basic protection, i.e.

416	▶ insulation of live parts
	▶ barriers or enclosures

      c fault protection

411	▶ automatic disconnection of supply, i.e.
	– presence of earthing conductor
	– presence of circuit protective conductors
	– presence of main protective bonding conductors
	– presence of supplementary bonding conductors (if required)
	– presence of adequate arrangements for alternative source(s), where applicable
	– FELV
	– choice and setting of protective and monitoring devices (for fault and/or overcurrent protection)
413	▶ electrical separation

      d additional protection by RCDs

411.3.3	
701.411.3.3	
515, 528	**ix** prevention of mutual detrimental influence (refer to 7.4)
132.15	**x** presence of appropriate devices for isolation and switching correctly located
	xi presence of undervoltage protective devices (where appropriate)
514	**xii** labelling of protective devices including circuit-breakers, RCDs, fuses, switches and terminals, main earthing and bonding connections
522	**xiii** selection of equipment and protective measures appropriate to external influences
132.12	**xiv** adequacy of access to switchgear and equipment
	xv presence of danger notices and other warning signs (see Section 6)
514.9.1	**xvi** presence of diagrams, instructions and similar information
522	**xvii** erection methods.

9.3 Testing 612

Testing must include the relevant tests from the following checklist.

When a test shows a failure to comply, the installation must be corrected. The test must then be repeated, as must any earlier test that could have been influenced by the failure. 612.1

9.3.1 Testing checklist

i   continuity of conductors:

    ▶ protective conductors including main and supplementary bonding conductors   612.2
    ▶ ring final circuit conductors including protective conductors

ii insulation resistance (between live conductors and between each live conductor and earth). Where appropriate during this measurement, line and neutral conductors may be connected together   612.3

iii polarity: this includes checks that single-pole control and protective devices (e.g. switches, circuit-breakers, fuses) are connected in the line conductor only, that bayonet and Edison-screw lampholders (except for E14 and E27 to BS EN 60238) have their outer contacts connected to the neutral conductor and that wiring has been correctly connected to socket-outlets and other accessories   612.6

iv earth electrode resistance (TT systems)   612.7
v   earth fault loop impedance (TN systems)   612.9
vi prospective short-circuit current and prospective earth fault current, if not determined by enquiry of the distributor   612.11
vii functional testing, including: 612.13

    ▶ testing of RCDs
    ▶ operation of all switchgear

viii verification of voltage drop (not normally required during initial verification).   612.14

Guidance on initial testing of installations

10.1 Safety and equipment

610.1
612.1

Electrical testing involves danger. It is the test operative's duty to ensure his or her own safety, and the safety of others, in the performance of the test procedures. When using test instruments, this is best achieved by precautions such as:

i an understanding of the correct application and use of the test instrumentation, leads, probes and accessories to be employed

ii checking that the test instrumentation is made in accordance with the appropriate safety standards such as BS EN 61243-3 for two-pole voltage detectors and BS EN 61010 or BS EN 61557 for instruments

iii checking before each use that all leads, probes, accessories (including all devices such as crocodile clips used to attach to conductors) and instruments including the proving unit are clean, undamaged and functioning; also, checking that isolation can be safely effected and that any locks or other means necessary for securing the isolation are available and functional

iv observing the safety measures and procedures set out in HSE Guidance Note GS 38 for all instruments, leads, probes and accessories. Some test instrument manufacturers advise that their instruments be used in conjunction with fused test leads and probes. Others advise the use of non-fused leads and probes when the instrument has in-built electrical protection, but it should be noted that such electrical protection does not extend to the probes and leads.

10.2 Sequence of tests

612.1

Note: The advice given does not preclude other test methods.

Tests should be carried out in the following sequence.

10.2.1 Before the supply is connected

i continuity of protective conductors, including main and supplementary bonding 612.2.1
ii continuity of ring final circuit conductors, including protective conductors 612.2.2
iii insulation resistance 612.3

612.6   **iv** polarity (by continuity methods)
612.7   **v** earth electrode resistance, using an earth electrode resistance tester (see **vii** also).

10.2.2 With the supply connected

  vi check polarity of supply, using an approved voltage indicator
  vii earth electrode resistance, using a loop impedance tester
612.8, 612.9   **viii** earth fault loop impedance
612.11   **ix** prospective fault current measurement, if not determined by enquiry of the distributor
612.13, 612.10   **x** functional testing, including RCDs and switchgear.

632.2   Results obtained during the various tests should be recorded on the Schedule of Test Results (Appendix 7) for future reference and checked for acceptability against prescribed criteria.

10.3 Test procedures

612.2.1   **10.3.1 Continuity of circuit protective conductors and protective bonding conductors (for ring final circuits see 10.3.2)**

Test methods 1 and 2 are alternative ways of testing the continuity of protective conductors.

Every protective conductor, including circuit protective conductors, the earthing conductor, main and supplementary bonding conductors, should be tested to verify that the conductors are electrically sound and correctly connected.

Test method 1 detailed below, as well as checking the continuity of the protective conductor, also measures (R_1 + R_2) which, when added to the external impedance (Z_e), enables the earth fault loop impedance (Z_s) to be checked **against the design**, see 10.3.6.

Note: (R_1 + R_2) is the sum of the resistances of the line conductor (R_1) and the circuit protective conductor (R_2) between the point of utilisation and origin of the installation.

Use an ohmmeter capable of measuring a low resistance for these tests.

Test method 1 can only be used to measure (R_1 + R_2) for an 'all insulated' installation, such as an installation wired in 'twin and earth'. Installations incorporating steel conduit, steel trunking, micc and pvc/swa cables will produce parallel paths to protective conductors. Such installations should be inspected for soundness of construction and test method 1 or 2 used to prove continuity.

10

i Continuity of circuit protective conductors

Continuity test method 1

Bridge the line conductor to the protective conductor at the distribution board so as to include all the circuit. Then test between line and earth terminals at each point in the circuit. The measurement at the circuit's extremity should be recorded and is the value of ($R_1 + R_2$) for the circuit under test (see Figure 10.1).

If the instrument does not include an 'auto-null' facility, or this is not used, the resistance of the test leads should be measured and deducted from the resistance readings obtained.

▼ **Figure 10.1** Connections for testing continuity of circuit protective conductors using test method 1

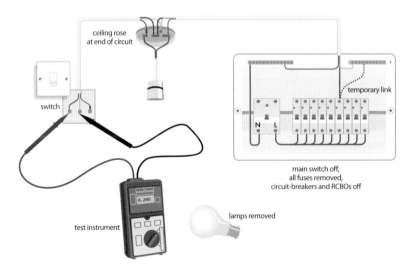

Continuity test method 2

Connect one terminal of the test instrument to a long test lead and connect this to the installation main earthing terminal.

Connect the other terminal of the instrument to another test lead and use this to make contact with the protective conductor at various points on the circuit, such as luminaires, switches, spur outlets, etc. (see Figure 10.2).

If the instrument does not include an 'auto-null' facility, or this is not used, the resistance of the test leads should be measured and deducted from the resistance readings obtained.

The resistance of the protective conductor R_2 is recorded on the Schedule of Test Results, Form 4 (see Appendix 7).

▼ **Figure 10.2**   Continuity test method 2

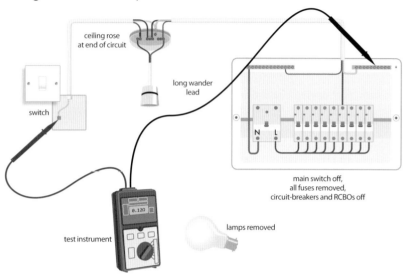

ii Continuity of the earthing conductor and protective bonding conductors

Continuity test method 2

For main bonding, connect one terminal of the test instrument to a long test lead and connect this to the installation main earthing terminal. Connect the other terminal of the instrument to another test lead and use this to make contact with the protective bonding conductor at its further end, such as at its connection to the incoming metal water, gas or oil service.

The connection verified boxes on the Electrical Installation Certificate should be ticked if the continuity of the earthing conductor and of each main bonding conductor is satisfactory, and the details of the material and the cross-sectional areas of the conductors recorded.

612.2.2   ## 10.3.2   Continuity of ring final circuit conductors

A three-step test is required to verify the continuity of the line, neutral and protective conductors and the correct wiring of a ring final circuit. The test results show if the ring has been interconnected to create an apparently continuous ring circuit which is in fact broken, or wrongly wired.

Use a low-resistance ohmmeter for this test.

Step 1

The line, neutral and protective conductors are identified at the distribution board and the end-to-end resistance of each is measured separately (see Figure 10.3). These resistances are r_1, r_n and r_2 respectively. A finite reading confirms that there is no open circuit on the ring conductors under test. The resistance values obtained should be the same (within 0.05 Ω) if the conductors are the same size. If the protective conductor has a reduced csa the resistance r_2 of the protective conductor loop will be proportionally higher than that of the line and neutral loops e.g. 1.67 times for 2.5/1.5 mm² cable. If these relationships are not achieved then either the conductors are incorrectly identified or there is something wrong at one or more of the accessories.

▼ **Figure 10.3** Step 1: The end-to-end resistances of the line, neutral and protective conductors are measured separately

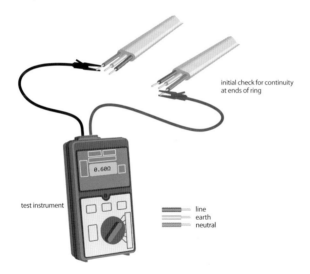

Step 2

The line and neutral conductors are then connected together at the distribution board so that the outgoing line conductor is connected to the returning neutral conductor and vice versa (see Figure 10.4). The resistance between line and neutral conductors is measured at each socket-outlet. The readings at each of the sockets wired into the ring will be substantially the same and the value will be approximately one-quarter of the resistance of the line plus the neutral loop resistances, i.e. $(r_1 + r_n)/4$. Any sockets wired as spurs will have a higher resistance value due to the resistance of the spur conductors.

Note: Where single-core cables are used, care should be taken to verify that the line and neutral conductors of opposite ends of the ring circuit are connected together. An error in this respect will be apparent from the readings taken at the socket-outlets, progressively increasing in value as readings are taken towards the midpoint of the ring, then decreasing again towards the other end of the ring.

▼ **Figure 10.4** Step 2: The line and neutral conductors are cross-connected and the resistance measured at each socket-outlet

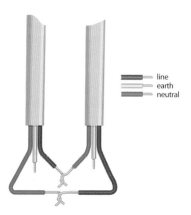

Step 3

The above step is then repeated, this time with the line and cpc cross-connected (see Figure 10.5). The resistance between line and earth is measured at each socket. The readings obtained at each of the sockets wired into the ring will be substantially the same and the value will be approximately one-quarter of the resistance of the line plus cpc loop resistances, i.e. $(r_1 + r_2)/4$. As before, a higher resistance value will be recorded at any sockets wired as spurs. The highest value recorded represents the maximum $(R_1 + R_2)$ of the circuit and is recorded on Form 4. The value can be used to determine the earth fault loop impedance (Z_s) of the circuit to verify compliance with the loop impedance requirements of BS 7671 (see 10.3.6).

▼ **Figure 10.5** Step 3: The line and cpc conductors are cross-connected and the resistance measured at each socket-outlet

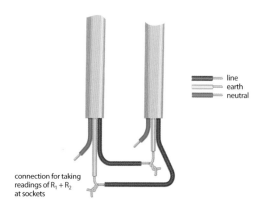

This sequence of tests also verifies the polarity of each socket, except that if the testing has been carried out at the terminals on the reverse of the accessories, a visual inspection is required to confirm correct polarity connections, and dispenses with the need for a separate polarity test.

10.3.3 Insulation resistance

i Pre-test checks

a Pilot or indicator lamps and capacitors are disconnected from circuits to prevent misleading test values from being obtained

b If a circuit includes voltage-sensitive electronic devices such as RCCBs, RCBOs or SRCDs incorporating electronic amplifiers, dimmer switches, touch switches, delay timers, power controllers, electronic starters or controlgear for fluorescent lamps, etc., either:

  1 the devices must be temporarily disconnected, or
  2 a measurement should be made between the live conductors (line and neutral) connected together and the protective earth only.

ii Tests

Tests should be carried out using the appropriate d.c. test voltage specified in Table 10.1.

The tests should be made at the distribution board with the main switch off, all fuses in place, switches and circuit-breakers closed, lamps removed and other current-using equipment disconnected. Where the removal of lamps and/or the disconnection of current-using equipment is impracticable, the local switches controlling such lamps and/or equipment should be open.

Where a circuit contains two-way switching, the two-way switches must be operated one at a time and further insulation resistance tests carried out to ensure that all the circuit wiring is tested.

▼ **Table 10.1** Minimum values of insulation resistance

Circuit nominal voltage	Test voltage (V d.c.)	Minimum insulation resistance (MΩ)
SELV and PELV	250	0.5
Up to and including 500 V with the exception of SELV and PELV, but including FELV	500	1.0

Notes:
1 Insulation resistance measurements are usually much higher than those of Table 10.1.
2 More stringent requirements are applicable for the wiring of fire alarm systems in buildings, see BS 5839-1.

For an installation operating at 230/400 V, although an insulation resistance value of only 1 MΩ complies with BS 7671, where an insulation resistance of less than, say, 2 MΩ is obtained the possibility of a latent defect exists. In these circumstances, each circuit should then be tested separately.

612.3.2 Where surge protective devices (SPDs) or other equipment such as electronic devices or RCDs with amplifiers are likely to influence the results of the test or may suffer damage from the test voltage, such equipment must be disconnected before carrying out the insulation resistance test.

Where it is not reasonably practicable to disconnect such equipment, the test voltage for the particular circuit may be reduced to 250 V d.c. but the insulation resistance must be at least 1 MΩ.

612.3.3 Where the circuit includes electronic devices which are likely to influence the results or be damaged, only a measurement between the live conductors connected together and earth should be made and the value should be not less than the value stated in Table 10.1.

iii Insulation resistance between live conductors

Single-phase and three-phase
Test between all the live (line and neutral) conductors at the distribution board (see Figure 10.6).

Resistance readings obtained should be not less than the minimum value referred to in Table 10.1.

▼ **Figure 10.6** Insulation resistance tests between live conductors of a circuit

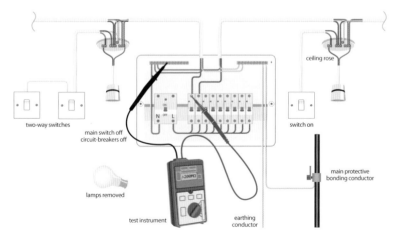

Note: The test may initially be carried out on the complete installation.

iv Insulation resistance to Earth

Single-phase

Test between the live conductors (line and neutral) and the circuit protective conductors at the distribution board (Figure 10.7 illustrates neutral to earth only).

For a circuit containing two-way switching or two-way and intermediate switching, the switches must be operated one at a time and the circuit subjected to additional insulation resistance tests.

▼ **Figure 10.7** Insulation resistance test between neutral and earth

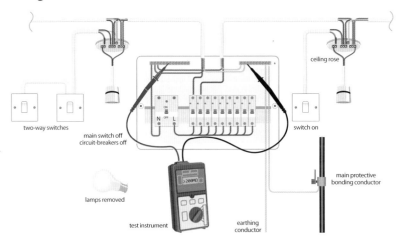

Notes:
1. The test may initially be carried out on the complete installation.
2. Earthing and bonding connections are in place.

Three-phase

Test to earth from all live conductors (including the neutral) connected together. Where a low reading is obtained it is necessary to test each conductor separately to earth, after disconnecting all equipment.

Resistance readings obtained should be not less than the minimum value referred to in Table 10.1.

v SELV and PELV circuits

Test between SELV and PELV circuits and live parts of other circuits at 500 V d.c.

612.4.1
612.4.2

Test between SELV or PELV conductors at 250 V d.c. and between PELV conductors and protective conductors of the PELV circuit at 250 V d.c.

vi FELV circuits

FELV circuits are tested as LV circuits at 500 V d.c.

10.3.4 Polarity

See Figure 10.8.

The method of test prior to connecting the supply is the same as test method 1 for checking the continuity of protective conductors which should have already been carried out (see 10.3.1). For ring final circuits a visual check may be required (see 10.3.2 following step 3).

It is important to confirm that:

1 overcurrent devices and single-pole controls are in the line conductor,
2 except for E14 and E27 lampholders to BS EN 60238, centre contact screw lampholders have the outer threaded contact connected to the neutral, and
3 socket-outlet polarities are correct.

After connection of the supply, polarity must be checked using a voltage indicator or a test lamp (in either case with leads complying with the recommendations of HSE Guidance Note GS 38).

▼ **Figure 10.8** Polarity test on a lighting circuit

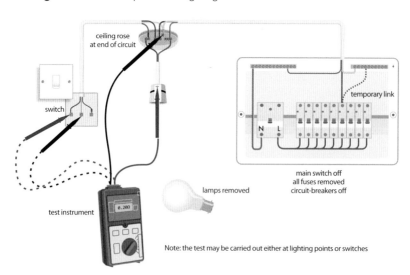

Note: the test may be carried out either at lighting points or switches

10.3.5 Earth electrode resistance

If the electrode under test is being used in conjunction with an RCD protecting an installation, the following method of test may be applied.

A loop impedance tester is connected between the line conductor at the origin of the installation and the earth electrode with the test link open, and a test performed. This impedance reading is treated as the electrode resistance and is then added to the resistance of the protective conductor for the protected circuits. The test should be carried out before energising the remainder of the installation.

The measured resistance should meet the following criteria and those of 10.3.6, but in any event should not exceed 200 Ω.

For TT systems, the value of the earth electrode resistance R_A in ohms multiplied by the operating current in amperes of the protective device $I_{\Delta n}$ should not exceed 50 V. For example, if $R_A = 200\ \Omega$, then the maximum RCD operating current should not exceed 250 mA.

411.5.3

REMEMBER TO REPLACE THE TEST LINK.

10.3.6 Earth fault loop impedance

612.9

The earth fault loop impedance (Z_s) is required to be determined for the furthest point of each circuit. It may be determined by

- direct measurement of Z_s,
- direct measurement of Z_e at the origin and adding ($R_1 + R_2$) measured during the continuity tests (10.3.1 and 10.3.2) {$Z_s = Z_e + (R1 + R_2)$},
- adding ($R_1 + R_2$) measured during the continuity tests to the value of Z_e declared by the distributor (see 1.1(iv) and 1.3(iv)). The effectiveness of the distributor's earth must be confirmed by a test.

The external impedance (Z_e) may be measured using a line-earth loop impedance tester.

The main switch is opened and made secure to isolate the installation from the source of supply. The earthing conductor is disconnected from the main earthing terminal and the measurement made between line and earth of the supply.

REMEMBER TO RECONNECT THE EARTHING CONDUCTOR TO THE EARTH TERMINAL AFTER THE TESTS.

Direct measurement of Z_s can only be made on a live installation. Neither the connection with earth nor bonding conductors are disconnected. The reading given by the loop impedance tester will usually be less than $Z_e + (R_1 + R_2)$ because of parallel earth return paths provided by any bonded extraneous-conductive-parts. This must be taken into account when comparing the results with design data.

Care should be taken to avoid any shock hazard to the testing personnel and to other persons on site during the tests.

The value of Z_s determined for each circuit should not exceed the value given in Appendix 2 for the particular overcurrent device and cable.

411.4.9 For TN systems, when protection is afforded by an RCD, the rated residual operating current in amperes times the earth fault loop impedance in ohms should not exceed 50 V. This test should be carried out before energising other parts of the system.

Note: For further information on the measurement of earth fault loop impedance, refer to Guidance Note 3 — *Inspection and Testing*.

10.3.7 Measurement of prospective fault current

612.11

It is not recommended that installation designs are based on measured values of prospective fault current, as changes to the distribution network subsequent to completion of the installation may increase fault levels.

Designs should be based on the maximum fault current provided by the distributor (see 7.2.7(i)).

If it is desired to measure prospective fault levels this should be done with all main bonding in place. Measurements are made at the distribution board between live conductors and between line conductors and earth.

For three-phase supplies, the maximum possible fault level will be approximately twice the single-phase to neutral value. (For three-phase to earth faults, neutral and earth path impedances have no influence.)

10.3.8 Check of phase sequence

612.12

In the case of three-phase circuits, it should be verified that the phase sequence is maintained.

10.3.9 Functional testing

612.13

RCDs should be tested as described in Section 11.

Switchgear, controls, etc., should be functionally tested; that is, operated to check that they work and are properly mounted and installed.

10.3.10 Verification of voltage drop

612.14

Note: Verification of voltage drop is not normally required during initial verification.

A new requirement has been introduced into BS 7671 that, where required, it should be verified that voltage drop does not exceed the limits stated in relevant product standards of installed equipment. Where no such limits are stated, voltage drop should be such that it does not impair the proper and safe functioning of installed equipment.

Typically, voltage drop will be evaluated using the measured circuit impedance.

The requirements for voltage drop are deemed to be met where the voltage drop between the origin and the relevant piece of equipment does not exceed the values stated in Appendix 12 of BS 7671:2008.

Appx 12

10

Appendix 12 gives maximum values of voltage drop for either lighting or other uses depending upon whether the installation is supplied directly from an LV distribution system or from a private LV supply.

It should be remembered that voltage drop may exceed the values stated in Appendix 12 in situations such as motor starting periods and where equipment has a high inrush current where such events remain within the limits specified in the relevant product standard or reasonable recommendation by a manufacturer.

Operation of RCDs   11

Residual current device (RCD) is the generic term for a device that operates when the residual current in the circuit reaches a predetermined value. An RCD is a protective device used to automatically disconnect the electrical supply when an imbalance is detected between the line and neutral conductors. In the case of a single-phase circuit, the device monitors the difference in currents between the line and neutral conductors. In a healthy circuit, where there is no earth fault current or protective conductor current, the sum of the currents in the line and neutral conductors is zero. If a line to earth fault develops, a portion of the line conductor current will not return through the neutral conductor. The device monitors this difference, operates and disconnects the circuit when the residual current reaches a preset limit, the residual operating current ($I_{\Delta n}$).

11.1 General test procedure

612.10

The tests are made on the load side of the RCD, as near as practicable to its point of installation and between the line conductor of the protected circuit and the associated circuit protective conductor. The load supplied should be disconnected during the test.

11.2 General purpose RCCBs to BS 4293

i   With a leakage current flowing equivalent to 50 per cent of the rated tripping current, the device should not open.

ii With a leakage current flowing equivalent to 100 per cent of the rated tripping current of the RCD, the device should open in less than 200 ms. Where the RCD incorporates an intentional time delay it should trip within a time range from '50% of the rated time delay plus 200 ms' to '100% of the rated time delay plus 200 ms'.

11.3 General purpose RCCBs to BS EN 61008 or RCBOs to BS EN 61009

i With a leakage current flowing equivalent to 50 per cent of the rated tripping current of the RCD, the device should not open.

ii With a leakage current flowing equivalent to 100 per cent of the rated tripping current of the RCD, the device should open in less than 300 ms unless it is of 'Type S' (or selective) which incorporates an intentional time delay. In this case, it should trip within a time range from 130 ms to 500 ms.

11.4 RCD protected socket-outlets to BS 7288

i With a leakage current flowing equivalent to 50 per cent of the rated tripping current of the RCD, the device should not open.

ii With a leakage current flowing equivalent to 100 per cent of the rated tripping current of the RCD, the device should open in less than 200 ms.

11.5 Additional protection

415.1.1 Where an RCD with a rated residual operating current $I_{\Delta n}$ not exceeding 30 mA is used to provide additional protection (against direct contact), with a test current of 5 $I_{\Delta n}$ the device should open in less than 40 ms. The maximum test time must not be longer than 40 ms, unless the protective conductor potential rises by less than 50 V. (The instrument supplier will advise on compliance.)

11.6 Integral test device

612.13.1 An integral test device is incorporated in each RCD. This device enables the electrical and mechanical parts of the RCD to be verified, by pressing the button marked 'T' or 'Test' (Figure 11.1).

Operation of the integral test device does not provide a means of checking:

a the continuity of the earthing conductor or the associated circuit protective conductors
b any earth electrode or other means of earthing
c any other part of the associated installation earthing.

The test button will only operate the RCD if the device is energised.

Confirm that the notice to test RCDs quarterly (by pressing the test button) is fixed in a prominent position (see 6.1(xi)).

▼ **Figure 11.1**   RCD operation

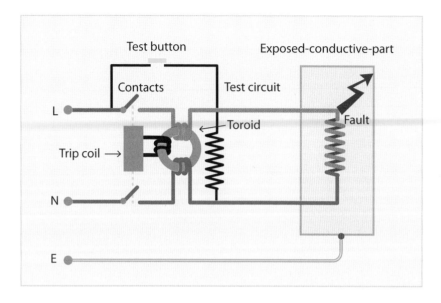

In Figure 11.1, a single-phase circuit, the device monitors the difference in currents between the line and neutral conductors. In a healthy circuit, where there is no earth fault current or protective conductor current, the sum of the currents in the line and neutral conductors is zero. If a line to earth fault develops, a portion of the line conductor current will not return through the neutral conductor. The device monitors this difference, operates and disconnects the circuit when the residual current reaches a preset limit, the residual operating current ($I_{\Delta n}$).

Appendix 1
Maximum demand and diversity

This appendix provides information on the determination of the maximum demand for an installation and includes the current demand to be assumed for commonly used equipment. It also includes some notes on the application of allowances for diversity.

The information and values given in this appendix are intended only for guidance because it is impossible to specify the appropriate allowances for diversity for every type of installation and such allowances call for special knowledge and experience. The values given in Table 1B, therefore, may be increased or decreased as decided by the installation designer concerned. No guidance is given for blocks of residential dwellings, large hotels, industrial and large commercial premises; such installations should be assessed on a case-by-case basis.

The current demand of a final circuit is determined by adding the current demands of all points of utilisation and equipment in the circuit and, where appropriate, making an allowance for diversity. Typical current demands to be used for this addition are given in Table 1A.

The current demand of an installation consisting of a number of final circuits may be assessed by using the allowances for diversity given in Table 1B which are applied to the total current demand of all the equipment supplied by the installation. The current demand of the installation should not be assessed by adding the current demands of the individual final circuits obtained as outlined above. In Table 1B the allowances are expressed either as percentages of the current demand or, where followed by the letters f.l. (full load), as percentages of the rated full load current of the current-using equipment. The current demand for any final circuit which is a standard circuit arrangement complying with Appendix 8 is the rated current of the overcurrent protective device of that circuit.

An alternative method of assessing the current demand of an installation supplying a number of final circuits is to add the diversified current demands of the individual circuits and then apply a further allowance for diversity. In this method the allowances given in Table 1B should not be used, the values to be chosen being the responsibility of the installation designer.

1 Appendix

The use of other methods of determining maximum demand is not precluded where specified by the installation designer. After the design currents for all the circuits have been determined, enabling the conductor sizes to be chosen, it is necessary to check that the limitation on voltage drop is met.

▼ **Table 1A** Current demand to be assumed for points of utilisation and current-using equipment

Point of utilisation or current-using equipment	Current demand to be assumed
Socket-outlets other than 2 A socket-outlets and other than 13 A socket-outlets See note 1	Rated current
2 A socket-outlets	At least 0.5 A
Lighting outlet See note 2	Current equivalent to the connected load, with a minimum of 100 W per lampholder
Electric clock, shaver supply unit (complying with BS EN 61558-2-5), shaver socket-outlet (complying with BS 4573), bell transformer, and current-using equipment of a rating not greater than 5 VA	May be neglected for the purpose of this assessment
Household cooking appliance	The first 10 A of the rated current plus 30% of the remainder of the rated current plus 5 A if a socket-outlet is incorporated in the control unit
All other stationary equipment	British Standard rated current, or normal current

Notes:
1 See Appendix 8 for the design of standard circuits using socket-outlets to BS 1363-2 and BS EN 60309-2 (BS 4343).
2 Final circuits for discharge lighting must be arranged so as to be capable of carrying the total steady current, viz. that of the lamp(s) and any associated controlgear and also their harmonic currents. Where more exact information is not available, the demand in volt-amperes is taken as the rated lamp watts multiplied by not less than 1.8. This multiplier is based upon the assumption that the circuit is corrected to a power factor of not less than 0.85 lagging, and takes into account controlgear losses and harmonic current.

Notes to Table 1B:
* In this context an instantaneous water-heater is considered to be a water-heater of any loading which heats water only while the tap is turned on and therefore uses electricity intermittently.
† It is important to ensure that distribution boards or consumer units are of sufficient rating to take the total load connected to them without the application of any diversity.

Appendix 1

Table 1B Allowances for diversity (see opposite for notes * and †)

Purpose of the final circuit fed from the conductors or switchgear to which the diversity applies	Individual household installations including individual dwellings of a block	Small shops, stores, offices and business premises	Small hotels, boarding houses, guest houses, etc.
1 Lighting	66% of total current demand	90% of total current demand	75% of total current demand
2 Heating and power (but see 3 to 8 below)	100% of total current demand up to 10 A +50% of any current demand in excess of 10 A	100% f.l. of largest appliance +75% f.l. of remaining appliances	100% f.l. of largest appliance +80% f.l. of second largest appliance +60% f.l. of remaining appliances
3 Cooking appliances	10 A +30% f.l. of connected cooking appliances in excess of 10 A +5 A if a socket-outlet is incorporated in the control unit	100% f.l. of largest appliance +80% f.l. of second largest appliance +60% f.l. of remaining appliances	100% f.l. of largest appliance +80% f.l. of second largest appliance +60% f.l. of remaining appliances
4 Motors (other than lift motors, which are subject to special consideration)	Not applicable	100% f.l. of largest motor +80% f.l. of second largest motor +60% f.l. of remaining motors	100% f.l. of largest motor +50% f.l. of remaining motors
5 Water-heaters (instantaneous type)*	100% f.l. of largest appliance +100% f.l. of second largest appliance +25% f.l. of remaining appliances	100% f.l. of largest appliance +100% f.l. of second largest appliance +25% f.l. of remaining appliances	100% f.l. of largest appliance +100% f.l. of second largest appliance +25% f.l. of remaining appliances
6 Water-heaters (thermostatically controlled)	No diversity allowable†		
7 Floor warming installations	No diversity allowable†		
8 Thermal storage space heating installations	No diversity allowable†		
9 Standard arrangement of final circuits in accordance with Appendix 8	100% of current demand of largest circuit +40% of current demand of every other circuit	100% of current demand of largest circuit +50% of current demand of every other circuit	
10 Socket-outlets other than those included in 9 above and stationary equipment other than those listed above	100% of current demand of largest point of utilisation +40% of current demand of every other point of utilisation	100% of current demand of largest point of utilisation +70% of current demand of every other point of utilisation	100% of current demand of largest point of utilisation +75% of current demand of every other point in main rooms (dining rooms, etc.) +40% of current demand of every other point of utilisation

Appendix 2
Maximum permissible measured earth fault loop impedance

The tables in this appendix provide maximum permissible measured earth fault loop impedances (Z_s) for compliance with BS 7671 where the standard final circuits of Table 7.1 are used. The values are those that must not be exceeded in the tests carried out under 10.3.6 at an ambient temperature of 10 °C. Table 2E provides correction factors for other ambient temperatures.

612.9
411.4.6
411.4.7
411.4.8

Where the cables to be used are to Table 4, 7 or 8 of BS 6004 or Table 3, 5, 6 or 7 of BS 7211 or are other thermoplastic (PVC) or thermosetting (low smoke halogen-free – LSHF) cables to these British Standards, and the cable loading is such that the maximum operating temperature is 70 °C, then Tables 2A–2C give the maximum earth fault loop impedances for circuits with:

1. protective conductors of copper and having from 1 mm² to 16 mm² cross-sectional area
2. an overcurrent protective device (i.e. a fuse) to BS 88 Part 2 or Part 6, BS 1361 or BS 3036.

For each type of fuse, two tables are given:

▶ where the circuit concerned is a final circuit not exceeding 32 A or a distribution circuit and the maximum disconnection time for compliance with Regulation 411.3.2.2 is 0.4 s for TN systems, and

411.3.2.2

▶ where the circuit concerned is a final circuit exceeding 32 A or a distribution circuit, and the disconnection time for compliance with Regulation 411.3.2.3 is 5 s for TN systems.

411.3.2.3

In each table the earth fault loop impedances given correspond to the appropriate disconnection time from a comparison of the time/current characteristic of the device concerned and the equation given in Regulation 543.1.3.

543.1.3

The tabulated values apply only when the nominal voltage to Earth (U_0) is 230 V.

Table 2D gives the maximum measured Z_s for circuits protected by circuit-breakers to BS 3871-1 and BS EN 60898, and RCBOs to BS EN 61009.

2 Appendix

Note: The impedances tabulated in this appendix are lower than those in Tables 41.2, 41.3 and 41.4 of BS 7671 as the impedances in this appendix are measured values at an assumed conductor temperature of 10 °C whilst those in BS 7671 are design figures at the conductor normal operating temperature. The correction factor (divisor) used is 1.24. For smaller section cables the impedance may also be limited by the adiabatic equation of Regulation 543.1.3. A value of k of 115 from Table 54.3 of BS 7671 is used. This is suitable for PVC insulated and sheathed cables to Table 4, 7 or 8 of BS 6004 and for lsf insulated and sheathed cables to Table 3, 5, 6 or 7 of BS 7211. The k value is based on both the thermoplastic (PVC) and thermosetting (LSHF) cables operating at a maximum temperature of 70 °C. The IEE Commentary on the Wiring Regulations provides a full explanation.

▼ **Table 2A** Semi-enclosed fuses. Maximum measured earth fault loop impedance (in ohms) at ambient temperature where the overcurrent protective device is a semi-enclosed fuse to BS 3036

i 0.4 second disconnection (final circuits in TN systems)

Protective conductor (mm²)	Fuse rating (A)			
	5	15	20	30
1.0	7.7	2.1	1.4	NP
≥ 1.5	7.7	2.1	1.4	0.9

ii 5 seconds disconnection (final circuits exceeding 32 A and distribution circuits in TN systems)

Protective conductor (mm²)	Fuse rating (A)			
	20	30	45	60
1.0	2.7	NP	NP	NP
1.5	3.1	2.0	NP	NP
2.5	3.1	2.1	1.2	NP
4.0	3.1	2.1	1.3	0.8
≥ 6.0	3.1	2.1	1.3	0.9

Note: NP means that the combination of the protective conductor and the fuse is Not Permitted.

Appendix 2

▼ **Table 2B** BS 88 fuses. Maximum measured earth fault loop impedance (in ohms) at ambient temperature where the overcurrent protective device is a fuse to BS 88

i 0.4 second disconnection (final circuits in TN systems)

Protective conductor (mm²)	Fuse rating (A)					
	6	10	16	20	25	32
1.0	6.9	4.1	2.2	1.4	1.2	0.66
1.5	6.9	4.1	2.2	1.4	1.2	0.84
≥ 2.5	6.9	4.1	2.2	1.4	1.2	0.84

ii 5 seconds disconnection (final circuits exceeding 32 A and distribution circuits in TN systems)

Protective conductor (mm²)	Fuse rating (A)							
	20	25	32	40	50	63	80	100
1.0	1.7	1.2	0.66	NP	NP	NP	NP	NP
1.5	2.3	1.7	1.1	0.64	NP	NP	NP	NP
2.5	2.3	1.8	1.5	0.93	0.55	0.34	NP	NP
4.0	2.3	1.8	1.5	1.1	0.77	0.50	0.23	NP
6.0	2.3	1.8	1.5	1.1	0.84	0.66	0.36	0.22
10.0	2.3	1.8	1.5	1.1	0.84	0.66	0.46	0.33
16.0	2.3	1.8	1.5	1.1	0.84	0.66	0.46	0.34

Note: NP means that the combination of the protective conductor and the fuse is Not Permitted.

2 Appendix

▼ **Table 2C** BS 1361 fuses. Maximum measured earth fault loop impedance (in ohms) at ambient temperature where the overcurrent protective device is a semi-enclosed fuse to BS 1361

i 0.4 second disconnection (final circuits in TN systems)

Protective conductor (mm²)	Fuse rating (A)			
	5	15	20	30
1.0	8.4	2.6	1.4	0.81
1.5	8.4	2.6	1.4	0.93
2.5 to 16	8.4	2.62	1.4	0.93

ii 5 seconds disconnection (final circuits exceeding 32 A and distribution circuits in TN systems)

Protective conductor (mm²)	Fuse rating (A)					
	20	30	45	60	80	100
1.0	1.7	0.81	NP	NP	NP	NP
1.5	2.2	1.2	0.34	NP	NP	NP
2.5	2.3	1.5	0.52	0.21	NP	NP
4.0	2.3	1.5	0.69	0.37	0.22	NP
6.0	2.3	1.5	0.77	0.53	0.30	0.15
10	2.3	1.5	0.77	0.56	0.40	0.22
16	2.3	1.5	0.77	0.56	0.40	0.29

Note: NP means that the combination of the protective conductor and the fuse is Not Permitted.

Appendix 2

▼ **Table 2D** Circuit-breakers. Maximum measured earth fault loop impedance (in ohms) at ambient temperature where the overcurrent device is a circuit-breaker to BS 3871 or BS EN 60898 or RCBO to BS EN 61009

0.1 to 5 second disconnection times

Circuit-breaker type	Circuit-breaker rating (A)												
	5	6	10	15	16	20	25	30	32	40	45	50	63	100
1	9.27	7.73	4.64	3.09	2.90	2.32	1.85	1.55	1.45	1.16	1.03	0.93	0.74	0.46
2	5.3	4.42	2.65	1.77	1.66	1.32	1.06	0.88	0.83	0.66	0.59	0.53	0.42	0.26
B	7.42	6.18	3.71	2.47	2.32	1.85	1.48	1.24	1.16	0.93	0.82	0.74	0.59	0.37
3&C	3.71	3.09	1.85	1.24	1.16	0.93	0.74	0.62	0.58	0.46	0.41	0.37	0.29	0.19
D	1.85	1.55	0.93	0.62	0.58	0.46	0.37	0.31	0.29	0.23	0.21	0.19	0.15	0.09

continues

2 Appendix

▼ **Table 2D**   continued

Minimum protective conductor size (mm²)*

Regulation 434.5.2 of BS 7671:2008 requires that the protective conductor csa meets the requirements of BS EN 60898-1, -2 or BS EN 61009-1, or the minimum quoted by the manufacturer. The values below are for energy limiting class 3, type B and C devices only.

Energy limiting class 3 device rating	Fault level (kA)	Protective conductor csa (mm²)	
		Type B	Type C
Up to and including 16 A	≤ 3	1.0	1.5
Up to and including 16 A	≤ 6	2.5	2.5
Over 16 up to and including 32 A	≤ 3	1.5	1.5
Over 16 up to and including 32 A	≤ 6	2.5	2.5
40 A	≤ 3	1.5	1.5
40 A	≤ 6	2.5	2.5

* For other device types and ratings or higher fault levels, consult manufacturer's data. See Regulation 434.5.2 and the IET publication *Commentary on the IEE Wiring Regulations*.

▼ **Table 2E**   Ambient temperature correction factors

Ambient temperature (°C)	Correction factor (from 10 °C) (notes 1 and 2)
0	0.96
5	0.98
10	1.00
20	1.04
25	1.06
30	1.08

Notes:
1 The correction factor is given by: {1 + 0.004(ambient temp − 10)} where 0.004 is the simplified resistance coefficient per °C at 20 °C given by BS EN 60228 for both copper and aluminium conductors.
2 The factors are different to those of Table 9B because Table 2E corrects from 10 °C and Table 9B from 20 °C.

The ambient correction factor of Table 2E is applied to the earth fault loop impedances of Tables 2A–D if the ambient temperature is other than 10 °C.

For example, if the ambient temperature is 25 °C the measured earth fault loop impedance of a circuit protected by a 32 A type B circuit-breaker to BS EN 60898 should not exceed 1.16 x 1.06 = 1.23 Ω.

# Appendix 3

Selection of types of cable and flexible cord for particular uses and external influences

For compliance with the requirements of Chapter 52 for the selection and erection of wiring systems in relation to risks of mechanical damage and corrosion, this lists, in two tables, types of cable and flexible cord suitable for the uses indicated. These tables are not intended to be exhaustive and other limitations may be imposed by the relevant regulations of BS 7671, in particular those concerning maximum permissible operating temperatures.

Chap 52

Information is also included in this appendix on protection against corrosion of exposed metalwork of wiring systems.

3 Appendix

▼ **Table 3A** Applications of cables for fixed wiring

Type of cable (note 7)	Uses	Comments
Thermoplastic (PVC) or thermosetting insulated non-sheathed cable (BS 7211, BS 7919)	For use in conduits, cable ducting or trunking	Intermediate support may be required on long vertical runs 70 °C maximum conductor temperature for normal wiring grades including thermosetting types (note 4) Cables run in PVC conduit should not operate with a conductor temperature greater than 70 °C (note 4)
Flat thermoplastic (PVC) or thermosetting insulated and sheathed cable (BS 6004)	For general indoor use in dry or damp locations. May be embedded in plaster	Additional mechanical protection may be necessary where exposed to mechanical stresses
	For use on exterior surface walls, boundary walls and the like	Protection from direct sunlight may be necessary. Black sheath colour is better for cables exposed to sunlight
	For use as overhead wiring between buildings	May need to be hard drawn (HD) copper conductors for overhead wiring (note 6)
	For use underground in conduits or pipes	Unsuitable for embedding directly in concrete
	For use in building voids or ducts formed in-situ	
Mineral insulated (BS EN 60702-1)	General	MI cables should have overall PVC covering where exposed to the weather or risk of corrosion, or where installed underground, or in concrete ducts
Thermoplastic or thermosetting insulated, armoured, thermoplastic sheathed (BS 5467, BS 6346, BS 6724, BS 7846)	General	Additional protection may be necessary where exposed to mechanical stresses Protection from direct sunlight may be necessary. Black sheath colour is better for cables exposed to sunlight

Notes:
1. The use of cable covers (preferably conforming to BS 2484) or equivalent mechanical protection is desirable for all underground cables which might otherwise subsequently be disturbed. Route marker tape should also be installed, buried just below ground level. Cables should be buried at a sufficient depth.
2. Cables having thermoplastic (PVC) insulation or sheath should preferably not be used where the ambient temperature is consistently below 0 °C or has been within the preceding 24 hours. Where they are to be installed during a period of low temperature, precautions should be taken to avoid risk of mechanical damage during handling. A minimum ambient temperature of 5 °C is advised in BS 7540:2005 (series) *Electric cables – Guide to use for cables with a rated voltage not exceeding 450/750 V* for some types of PVC insulated and sheathed cables.
3. Cables must be suitable for the maximum ambient temperature, and must be protected from any excess heat produced by other equipment, including other cables.

Appendix 3

4   Thermosetting cable types (to BS 7211 or BS 5467) can operate with a conductor temperature of 90 °C. This must be limited to 70 °C where drawn into a conduit, etc., with thermoplastic (PVC) insulated conductors or connected to electrical equipment (512.1.2 and 523.1), or where such cables are installed in plastic conduit or trunking.

5   For cables to BS 6004, BS 6007, BS 7211, BS 6346, BS 5467 and BS 6724, further guidance may be obtained from those standards. Additional advice is given in BS 7540:2005 (series) *Guide to use of cables with a rated voltage not exceeding 450/750 V* for cables to BS 6004, BS 6007 and BS 7211.

6   Cables for overhead wiring between buildings must be able to support their own weight and any imposed wind or ice/snow loading. A catenary support is usual but hard drawn copper types may be used.

7   **BS 5467: Electric cables.** Thermosetting insulated, armoured cables for voltages of 600/1000 V and 1900/3300 V

BS 6004: Electric cables. PVC insulated, non-armoured cables for voltages up to and including 450/750 V for electric power, lighting and internal wiring

BS 6346: Electric cables. PVC insulated, armoured cables for voltages of 600/1000 V and 1900/3300 V

BS 6724: Electric cables. Thermosetting insulated, armoured cables for voltages of 600/1000 V and 1900/3300 V, having low emission of smoke and corrosive gases when affected by fire

BS 7211: Electric cables. Thermosetting insulated, non-armoured cables for voltages up to and including 450/750 V, for electric power, lighting and internal wiring, and having low emission of smoke and corrosive gases when affected by fire

BS 7846: Electric cables. 600/1000 V armoured fire-resistant cables having thermosetting insulation and low emission of smoke and corrosive gases when affected by fire

BS EN 60702-1: Mineral insulated cables and their terminations with a rated voltage not exceeding 750 V. Cables

Migration of plasticiser from thermoplastic (PVC) materials

Thermoplastic (PVC) sheathed cables, including thermosetting insulated with thermoplastic sheath, e.g. LSHF, must be separated from expanded polystyrene materials to prevent take-up of the cable plasticiser by the polystyrene as this will reduce the flexibility of the cables.

Thermal insulation

Thermoplastic (PVC) sheathed cables in roof spaces must be clipped clear of any insulation made of expanded polystyrene granules.

Cable clips

Polystyrene cable clips are softened by contact with thermoplastic (PVC). Nylon and polypropylene are unaffected.

Grommets

Natural rubber grommets can be softened by contact with thermoplastic (PVC). Synthetic rubbers are more resistant. Thermoplastic (PVC) grommets are not affected, but could affect other plastics.

3 Appendix

Wood preservatives

Thermoplastic (PVC) sheathed cables should be covered to prevent contact with preservative fluids during application. After the solvent has evaporated (good ventilation is necessary) the preservative has no effect.

Creosote

Creosote should not be applied to thermoplastic (PVC) sheathed cables because it causes decomposition, solution, swelling and loss of pliability.

▼ **Table 3B**   Applications of flexible cables and cords to BS 6500:2000 and BS 7919:2001

Type of flexible cord	Uses
Light thermoplastic (PVC) insulated and sheathed flexible cord	Indoors in household or commercial premises in dry situations, for light duty
Ordinary thermoplastic (PVC) insulated and sheathed flexible cord	Indoors in household or commercial premises, including damp situations, for medium duty
	For cooking and heating appliances where not in contact with hot parts
	For outdoor use other than in agricultural or industrial applications
	For electrically powered hand tools
60 °C thermosetting (rubber) insulated braided twin and three-core flexible cord	Indoors in household or commercial premises where subject only to low mechanical stresses
60 °C thermosetting (rubber) insulated and sheathed flexible cord	Indoors in household or commercial premises where subject only to low mechanical stresses
	For occasional use outdoors
	For electrically powered hand tools
60 °C thermosetting (rubber) insulated oil-resisting with flame-retardant sheath	For general use, unless subject to severe mechanical stresses
	For use in fixed installations where protected by conduit or other enclosure
90 °C thermosetting (rubber) insulated HOFR sheathed	General, including hot situations, e.g. night storage heaters, immersion heaters and boilers
90 °C heat-resisting thermoplastic (PVC) insulated and sheathed	General, including hot situations, e.g. for pendant luminaires
150 °C thermosetting (rubber) insulated and braided	For use at high ambient temperatures
	For use in or on luminaires
185 °C glass-fibre insulated single-core, twisted twin and three-core	For internal wiring of luminaires only and then only where permitted by BS 4533
185 °C glass-fibre insulated braided circular	For dry situations at high ambient temperatures and not subject to abrasion or undue flexing
	For the wiring of luminaires

Appendix 3

Notes to Table 3B:
1. Cables and cords having thermoplastic (PVC) insulation or sheath should preferably not be used where the ambient temperature is consistently below 0 °C. Where they are to be installed during a period of low temperature, precautions should be taken to avoid risk of mechanical damage during handling.
2. Cables and cords should be suitable for the maximum ambient temperature, and should be protected from any excess heat produced by other equipment, including other cables.
3. For flexible cords and cables to BS 6007, BS 6141 and BS 6500 further guidance may be obtained from those standards, or from BS 7540:2005 (series) *Guide to use of cables with a rated voltage not exceeding 450/750 V*.
4. Where used as connections to equipment, flexible cables and cords should, where possible, be of the minimum practicable length to minimize danger. The length of the flexible cable or cord must be such that will permit correct operation of the protective device.
5. Where attached to equipment flexible cables and cords should be protected against tension, crushing, abrasion, torsion and kinking, particularly at the inlet point to the electrical equipment. At such inlet points it may be necessary to use a device which ensures that the cable is not bent to an internal radius below that given in the appropriate part of Table 4 of BS 6700. Strain relief, clamping devices or cord guards should not damage the cord.
6. Flexible cables and cords should not be run under carpets or other floor coverings where furniture or other equipment may rest on them or where heat dissipation from the cable will be affected. Flexible cables and cords should not be placed where there is a risk of damage from traffic passing over them, unless suitably protected.
7. Flexible cables and cords should not be used in contact with or close to heated surfaces, especially if the surface approaches the upper thermal limit of the cable or cord.

Protection of exposed metalwork and wiring systems against corrosion

522.3
522.5

In damp situations, where metal cable sheaths and armour of cables, metal conduit and conduit fittings, metal ducting and trunking systems, and associated metal fixings, are liable to chemical deterioration or electrolytic attack by materials of a structure with which they may come in contact, it is necessary to take suitable precautions against corrosion.

Materials likely to cause such attack include:

- materials containing magnesium chloride which are used in the construction of floors and plaster mouldings,
- plaster undercoats which may include corrosive salts,
- lime, cement and plaster, for example on unpainted walls,
- oak and other acidic woods,
- dissimilar metals likely to set up electrolytic action.

Application of suitable coatings before erection, or prevention of contact by separation with plastics, are recognized as effective precautions against corrosion.

Special care is required in the choice of materials for clips and other fittings for bare aluminium sheathed cables and for aluminium conduit, to avoid risk of local corrosion in damp situations. Examples of suitable materials for this purpose are the following:

522.5.2
522.5.3

- porcelain,
- plastics,
- aluminium;
- corrosion-resistant aluminium alloys,

3 | Appendix

- zinc alloys complying with BS 1004,
- iron or steel protected against corrosion by galvanizing, sherardizing, etc.

522.5.2 Contact between bare aluminium sheaths or aluminium conduits and any parts made of brass or other metal having a high copper content should be especially avoided in damp situations, unless the parts are suitably plated. If such contact is unavoidable, the joint should be completely protected against ingress of moisture. Wiped joints in aluminium sheathed cables should always be protected against moisture by a suitable paint, by an impervious tape, or by embedding in bitumen.

Appendix 4
Methods of support for cables, conductors and wiring systems

This appendix describes examples of methods of support for cables, conductors and wiring systems which should satisfy the relevant requirements of Chapter 52 of BS 7671. The use of other methods is not precluded where specified by a suitably qualified electrical engineer.

522.8

Cables generally

Items 1 to 8 below are generally applicable to supports on structures which are subject only to vibration of low severity and a low risk of mechanical impact.

1. For non-sheathed cables, installation in conduit without further fixing of the cables, precautions being taken against undue compression or other mechanical stressing of the insulation at the top of any vertical runs exceeding 5 m in length.
2. For cables of any type, installation in ducting or trunking without further fixing of the cables, vertical runs not exceeding 5 m in length without intermediate support.
3. For sheathed and/or armoured cables installed in accessible positions, support by clips at spacings not exceeding the appropriate value stated in Table 4A.
4. For cables of any type, resting without fixing in horizontal runs of ducts, conduits, cable ducting or trunking.
5. For sheathed and/or armoured cables in horizontal runs which are inaccessible and unlikely to be disturbed, resting without fixing on part of a building, the surface of that part being reasonably smooth.
6. For sheathed-and-armoured cables in vertical runs which are inaccessible and unlikely to be disturbed, supported at the top of the run by a clip and a rounded support of a radius not less than the appropriate value stated in Table 4E.
7. For sheathed cables without armour in vertical runs which are inaccessible and unlikely to be disturbed, supported by the method described in Item 6 above; the length of run without intermediate support not exceeding 5 m for a thermosetting or thermoplastic sheathed cable.

4 Appendix

8   For thermosetting or thermoplastic (PVC) sheathed cables, installation in conduit without further fixing of the cables, any vertical runs being in conduit of suitable size and not exceeding 5 m in length.

Particular applications

721.522.8.1.3 9   In caravans, for sheathed cables in inaccessible spaces such as ceiling, wall and floor spaces, support at intervals not exceeding 0.25 m for horizontal runs and 0.4 m for vertical runs.

10 In caravans, for horizontal runs of sheathed cables passing through floor or ceiling joists in inaccessible floor or ceiling spaces, securely bedded in thermal insulating material, no further fixing is required.

559.6.1.3
559.6.1.5 11 For flexible cords used as pendants, attachment to a ceiling rose or similar accessory by the cord grip or other method of strain relief provided in the accessory.

12 For temporary installations and installations on construction sites, supports so arranged that there is no appreciable mechanical strain on any cable termination or joint.

Overhead wiring

13 For cables sheathed with thermosetting or thermoplastic material, supported by a separate catenary wire, either continuously bound up with the cable or attached thereto at intervals, the intervals not exceeding those stated in column 2 of Table 4A.

14 Support by a catenary wire incorporated in the cable during manufacture, the spacings between supports not exceeding those stated by the manufacturer and the minimum height above ground being in accordance with Table 4B.

15 For spans without intermediate support (e.g. between buildings) of thermoplastic (PVC) insulated thermoplastic (PVC) sheathed cable, or thermosetting insulated cable having an oil-resisting and flame-retardant or HOFR sheath, terminal supports so arranged that:

▶ no undue strain is placed upon the conductors or insulation of the cable,
▶ adequate precautions are taken against any risk of chafing of the cable sheath, and
▶ the minimum height above ground and the length of such spans are in accordance with the appropriate values indicated in Table 4B.

16 Bare or thermoplastic (PVC) covered conductors of an overhead line for distribution between a building and a remote point of utilisation (e.g. another building) supported on insulators, the lengths of span and heights above ground having the appropriate values indicated in Table 4B or otherwise installed in accordance with the Electricity Safety, Quality and Continuity Regulations 2002 (as amended).

17 For spans without intermediate support (e.g. between buildings) and which are in situations inaccessible to vehicular traffic, cables installed in heavy gauge steel conduit, the length of span and height above ground being in accordance with Table 4B.

Appendix 4

Conduit and cable trunking

18 Rigid conduit supported in accordance with Table 4C.
19 Cable trunking supported in accordance with Table 4D.
20 Conduit embedded in the material of the building.
21 Pliable conduit embedded in the material of the building or in the ground, or supported in accordance with Table 4C.

▼ **Table 4A** Spacings of supports for cables in accessible positions

Overall diameter of cable, d* (mm)	Maximum spacings of clips (mm)							
	Non-armoured thermosetting or thermoplastic (PVC) sheathed cables				Armoured cables		Mineral insulated copper sheathed or aluminium sheathed cables	
	Generally		In caravans					
	Horizontal †	Vertical †	Horizontal †	Vertical †	Horizontal †	Vertical †	Horizontal †	Vertical †
1	2	3	4	5	6	7	8	9
d ≤ 9	250	400	250 (for all sizes)	400 (for all sizes)	–	–	600	800
9 < d ≤ 15	300	400			350	450	900	1200
15 < d ≤ 20	350	450			400	550	1500	2000
20 < d ≤ 40	400	550			450	600	–	–

Note: For the spacing of supports for cables having an overall diameter exceeding 40 mm, the manufacturer's recommendations should be observed.

* For flat cables taken as the dimension of the major axis.
† The spacings stated for horizontal runs may be applied also to runs at an angle of more than 30° from the vertical. For runs at an angle of 30° or less from the vertical, the vertical spacings are applicable.

4 Appendix

▼ **Table 4B** Maximum lengths of span and minimum heights above ground for overhead wiring between buildings, etc.

Type of system	Maximum length of span (m)	Minimum height of span above ground (m)†		
		At road crossings	In positions accessible to vehicular traffic, other than crossings	In positions inaccessible to vehicular traffic*
1	2	3	4	5
Cables sheathed with thermoplastic (PVC) or having an oil-resisting and flame-retardant or HOFR sheath, without intermediate support.	3	5.8	5.8	3.5
Cables sheathed with thermoplastic (PVC) or having an oil-resisting and flame-retardant or HOFR sheath, in heavy gauge steel conduit of diameter not less than 20 mm and not jointed in its span.	3	5.8	5.8	3
Thermoplastic (PVC) covered overhead lines on insulators without intermediate support.	30	5.8	5.8	3.5
Bare overhead lines on insulators without intermediate support.	30	5.8	5.8	5.2
Cables sheathed with thermoplastic (PVC) or having an oil-resisting and flame-retardant or HOFR sheath, supported by a catenary wire.	No limit	5.8	5.8	3.5
Aerial cables incorporating a catenary wire.	Subject to Item 14	5.8	5.8	3.5

417.3 A bare or insulated overhead line for distribution between buildings and structures must be installed to the standard required by the Electricity Safety, Quality and Continuity Regulations 2002.

* Column 5 is not applicable in agricultural premises.
† In some special cases, such as where cranes are present, it will be necessary to increase the minimum height of span above ground. It is preferable to use underground cables in such locations.

Appendix 4

▼ **Table 4C**   Spacings of supports for conduits

Nominal diameter of conduit, d (mm)	Maximum distance between supports (m)					
	Rigid metal		Rigid insulating		Pliable	
	Horizontal	Vertical	Horizontal	Vertical	Horizontal	Vertical
1	2	3	4	5	6	7
$d \leq 16$	0.75	1.0	0.75	1.0	0.3	0.5
$16 < d \leq 25$	1.75	2.0	1.5	1.75	0.4	0.6
$25 < d \leq 40$	2.0	2.25	1.75	2.0	0.6	0.8
$d > 40$	2.25	2.5	2.0	2.0	0.8	1.0

Notes:
1 The spacings tabulated allow for maximum fill of cables permitted by the Regulations and the thermal limits specified in the relevant British Standards. They assume that the conduit is not exposed to other mechanical stress.
2 Supports should be positioned within 300 mm of bends or fittings. A flexible conduit should be of such length that it does not need to be supported in its run.
3 The inner radius of a conduit bend should be not less than 2.5 times the outside diameter of the conduit.

▼ **Table 4D**   Spacings of supports for cable trunking

Cross-sectional area of trunking, A (mm²)	Maximum distance between supports (m)			
	Metal		Insulating	
	Horizontal	Vertical	Horizontal	Vertical
1	2	3	4	5
$300 < A \leq 700$	0.75	1.0	0.5	0.5
$700 < A \leq 1500$	1.25	1.5	0.5	0.5
$1500 < A \leq 2500$	1.75	2.0	1.25	1.25
$2500 < A \leq 5000$	3.0	3.0	1.5	2.0
$A > 5000$	3.0	3.0	1.75	2.0

Notes:
1 The spacings tabulated allow for maximum fill of cables permitted by the Regulations and the thermal limits specified in the relevant British Standards. They assume that the trunking is not exposed to other mechanical stress.
2 The above figures do not apply to lighting suspension trunking, where the manufacturer's instructions must be followed, or where special strengthening couplers are used. Supports should be positioned within 300 mm of bends or fittings.

4 Appendix

▼ **Table 4E** Minimum internal radii of bends in cables for fixed wiring

Insulation	Finish	Overall diameter, d* (mm)	Factor to be applied to overall diameter of cable to determine minimum internal radius of bend
Thermosetting or thermoplastic (PVC) (circular, or circular stranded copper or aluminium conductors)	Non-armoured	d ≤ 10 10 < d ≤ 25 d > 25	3(2)† 4(3)† 6
	Armoured	Any	6
Thermosetting or thermoplastic (PVC) (solid aluminium or shaped copper conductors)	Armoured or non-armoured	Any	8
Mineral	Copper sheath with or without covering	Any	6‡

* For flat cables the diameter refers to the major axis.
† The value in brackets relates to single-core circular conductors of stranded construction installed in conduit, ducting or trunking.
‡ Mineral insulated cables may be bent to a radius not less than three times the cable diameter over the copper sheath, provided that the bend is not reworked, i.e. straightened and re-bent.

Appendix 5
Cable capacities of conduit and trunking

A number of variable factors affect any attempt to arrive at a standard method of assessing the capacity of conduit or trunking.

522.8.2
522.8.3

Some of these are:

- reasonable care (of drawing-in)
- acceptable use of the space available
- tolerance in cable sizes
- tolerance in conduit and trunking.

The following tables can only give guidance on the maximum number of cables which should be drawn in. The sizes should ensure an easy pull with low risk of damage to the cables.

Only the ease of drawing-in is taken into account. The electrical effects of grouping are not. As the number of circuits increases the installed current-carrying capacity of the cable decreases. Cable sizes have to be increased with consequent increase in cost of cable and conduit.

It may sometimes be more attractive economically to divide the circuits concerned between two or more enclosures.

If thermosetting cables are installed in the same conduit or trunking as thermoplastic (PVC) insulated cables, the conductor operating temperature of any of the cables must not exceed that for thermoplastic (PVC), i.e. thermosetting cables must be rated as thermoplastic (PVC).

The following three cases are dealt with. Single-core thermoplastic (PVC) insulated cables in:

   i straight runs of conduit not exceeding 3 m in length (Tables 5A and 5B).
   ii straight runs of conduit exceeding 3 m in length, or in runs of any length incorporating bends or sets (Tables 5C and 5D).
   iii trunking (Tables 5E and 5F).

For cables and/or conduits not covered by this appendix, advice on the number of cables that can be drawn in should be obtained from the manufacturer.

5 Appendix

i Single-core thermoplastic (PVC) insulated cables in straight runs of conduit not exceeding 3 m in length

For each cable it is intended to use, obtain the appropriate factor from Table 5A.

Add the cable factors together and compare the total with the conduit factors given in Table 5B.

The minimum conduit size is that having a factor equal to or greater than the sum of the cable factors.

▼ **Table 5A**
Cable factors for use in conduit in short straight runs

Type of conductor	Conductor cross-sectional area (mm²)	Cable factor
Solid	1	22
	1.5	27
	2.5	39
Stranded	1.5	31
	2.5	43
	4	58
	6	88
	10	146
	16	202
	25	385

▼ **Table 5B**
Conduit factors for use in short straight runs

Conduit diameter (mm)	Conduit factor
16	290
20	460
25	800
32	1400
38	1900
50	3500
63	5600

ii Single-core thermoplastic (PVC) insulated cables in straight runs of conduit exceeding 3 m in length, or in runs of any length incorporating bends or sets

For each cable it is intended to use, obtain the appropriate factor from Table 5C.

Add the cable factors together and compare the total with the conduit factors given in Table 5D, taking into account the length of run it is intended to use and the number of bends and sets in that run.

The minimum conduit size is that size having a factor equal to or greater than the sum of the cable factors. For the larger sizes of conduit, multiplication factors are given relating them to 32 mm diameter conduit.

Appendix 5

▼ **Table 5C** Cable factors for use in conduit in long straight runs over 3 m, or runs of any length incorporating bends

Type of conductor	Conductor cross-sectional area (mm²)	Cable factor
Solid or Stranded	1	16
	1.5	22
	2.5	30
	4	43
	6	58
	10	105
	16	145
	25	217

The inner radius of a conduit bend should be not less than 2.5 times the outside diameter of the conduit.

5 Appendix

▼ **Table 5D** Cable factors for runs incorporating bends and long straight runs

Length of run (m)	Conduit diameter (mm)																
	Straight				One bend				Two bends				Three bends				
	16	20	25	32	16	20	25	32	16	20	25	32	16	20	25	32	
1	Covered by Tables 5A and 5B				188	303	543	947	177	286	514	900	158	256	463	818	
1.5					182	294	528	923	167	270	487	857	143	233	422	750	
2					177	286	514	900	158	256	463	818	130	213	388	692	
2.5					171	278	500	878	150	244	442	783	120	196	358	643	
3					167	270	487	857	143	233	422	750	111	182	333	600	
3.5	179	290	521	911	162	263	475	837	136	222	404	720	103	169	311	563	
4	177	286	514	900	158	256	463	818	130	213	388	692	97	159	292	529	
4.5	174	282	507	889	154	250	452	800	125	204	373	667	91	149	275	500	
5	171	278	500	878	150	244	442	783	120	196	358	643	86	141	260	474	
6	167	270	487	857	143	233	422	750	111	182	333	600					
7	162	263	475	837	136	222	404	720	103	169	311	563					
8	158	256	463	818	130	213	388	692	97	159	292	529					
9	154	250	452	800	125	204	373	667	91	149	275	500					
10	150	244	442	783	120	196	358	643	86	141	260	474					

Length of run (m)	Four bends			
	16	20	25	32
1	130	213	388	692
1.5	111	182	333	600
2	97	159	292	529
2.5	86	141	260	474

Additional factors:

▲ For 38 mm diameter use 1.4 × (32 mm factor)
▲ For 50 mm diameter use 2.6 × (32 mm factor)
▲ For 63 mm diameter use 4.2 × (32 mm factor)

Appendix 5

iii Single-core thermoplastic (PVC) insulated cables in trunking

For each cable it is intended to use, obtain the appropriate factor from Table 5E.

Add the cable factors together and compare the total with the factors for trunking given in Table 5F.

The minimum size of trunking is that size having a factor equal to or greater than the sum of the cable factors.

▼ **Table 5E**   Cable factors for trunking

Type of conductor	Conductor cross-sectional area (mm²)	PVC BS 6004 Cable factor	Thermosetting BS 7211 Cable factor
Solid	1.5	8.0	8.6
	2.5	11.9	11.9
Stranded	1.5	8.6	9.6
	2.5	12.6	13.9
	4	16.6	18.1
	6	21.2	22.9
	10	35.3	36.3
	16	47.8	50.3
	25	73.9	75.4

Notes:
1 These factors are for metal trunking and may be optimistic for plastic trunking, where the cross-sectional area available may be significantly reduced from the nominal by the thickness of the wall material.
2 The provision of spare space is advisable; however, any circuits added at a later date must take into account grouping, Regulation 523.5.

5 Appendix

▼ **Table 5F** Factors for trunking

Dimensions of trunking (mm x mm)	Factor	Dimensions of trunking (mm x mm)	Factor
50 x 38	767	200 x 100	8572
50 x 50	1037	200 x 150	13001
75 x 25	738	200 x 200	17429
75 x 38	1146	225 x 38	3474
75 x 50	1555	225 x 50	4671
75 x 75	2371	225 x 75	7167
100 x 25	993	225 x 100	9662
100 x 38	1542	225 x 150	14652
100 x 50	2091	225 x 200	19643
100 x 75	3189	225 x 225	22138
100 x 100	4252	300 x 38	4648
150 x 38	2999	300 x 50	6251
150 x 50	3091	300 x 75	9590
150 x 75	4743	300 x 100	12929
150 x 100	6394	300 x 150	19607
150 x 150	9697	300 x 200	26285
200 x 38	3082	300 x 225	29624
200 x 50	4145	300 x 300	39428
200 x 75	6359		

Note: Space factor is 45% with trunking thickness taken into account.

Other sizes and types of cable or trunking

For sizes and types of cable or trunking other than those given in Tables 5E and 5F, the number of cables installed should be such that the resulting space factor does not exceed 45 per cent of the net internal cross-sectional area.

Space factor is the ratio (expressed as a percentage) of the sum of the overall cross-sectional areas of cables (including insulation and any sheath) to the internal cross-sectional area of the trunking or other cable enclosure in which they are installed. The effective overall cross-sectional area of a non-circular cable is taken as that of a circle of diameter equal to the major axis of the cable.

Care should be taken to use trunking bends etc which do not impose bending radii on cables less than those required by Table 4E.

Appendix 6
Current-carrying capacities and voltage drop for copper conductors

Current-carrying capacity

In this simplified approach the assumption is made that the overcurrent protective device provides both fault current and overload current protection.

Ch 52
Appx 4
435.1

Procedure

1 The design current I_b of the circuit must first be established.
2 The overcurrent device rating I_n is then selected so that I_n is greater than or equal to I_b

Appx 4,3
433.1.1

$$I_n \geq I_b$$

The tabulated current-carrying capacity of the selected cable I_t is then given by

$$I_t \geq \frac{I_n}{C_a \, C_g \, C_i \, C_c}$$

for simultaneously occurring factors.

C is a rating factor to be applied where the installation conditions differ from those for which values of current-carrying capacity are tabulated in this appendix. The various rating factors are identified as follows:

Appx 4,3

C_a for ambient temperature, see Table 6A
C_g for grouping, see Table 6C
C_i for thermal insulation, see Table 6B (Note: For cables installed in thermal insulation as described in Tables 6D1, 6E1 and 6F, $C_i = 1$)

6 | Appendix

C_c for the type of protective device or installation condition, i.e.:
- where the protective device is a semi-enclosed fuse to BS 3036, $C_c = 0.725$
- where the cable installation method is 'in a duct in the ground' or 'buried direct', $C_c = 0.9$
- if both the above apply, $C_c = 0.725 \times 0.9 = 0.653$
- for all other cases $C_c = 1$

Voltage drop

To calculate the voltage drop in volts the tabulated value of voltage drop (mV/A/m) has to be multiplied by the design current of the circuit (I_b), the length of run in metres (L), and divided by 1000 (to convert to volts)

$$\text{voltage drop} = \frac{(mV/A/m) \times I_b \times L}{1000}$$

The requirements of BS 7671 are deemed to be satisfied if the voltage drop between the origin of the installation and a lighting point does not exceed 3 per cent of the nominal voltage (6.9 V) and for other current-using equipment or socket-outlets does not exceed 5 per cent (11.5 V single-phase).

▼ **Table 6A** Rating factors for ambient air temperatures other than 30 °C to be applied to the current-carrying capacities for cables in free air

Ambient temperature (°C)	Insulation			
			Mineral	
	70 °C thermoplastic	90 °C thermosetting	Thermoplastic covered or bare and exposed to touch 70 °C	Bare and not exposed to touch 105 °C
25	1.03	1.02	1.07	1.04
30	1.00	1.00	1.00	1.00
35	0.94	0.96	0.93	0.96
40	0.87	0.91	0.85	0.92

Appendix 6

Thermal insulation

523.7

Where a cable is to be run in a space to which thermal insulation is likely to be applied, the cable should, wherever practicable, be fixed in a position such that it will not be covered by the thermal insulation. Where fixing in such a position is impracticable, the cross-sectional area of the cable must be increased appropriately.

For a cable installed in thermal insulation as described in Tables 6D1, 6E1 and 6F no correction is required.

Note: Reference methods 100, 101 and 102 require the cable to be in contact with the plasterboard or the joists, see Tables 7.2 and 7.3 in Section 7.

For a single cable likely to be totally surrounded by thermally insulating material over a length of more than 0.5 m, the current-carrying capacity should be taken, in the absence of more precise information, as 0.5 times the current-carrying capacity for that cable clipped direct to a surface and open (reference method C).

Where a cable is totally surrounded by thermal insulation for less than 0.5 m the current-carrying capacity of the cable should be reduced appropriately depending on the size of cable, length in insulation and thermal properties of the insulation. The derating factors in Table 6B are appropriate to conductor sizes up to 10 mm^2 in thermal insulation having a thermal conductivity (λ) greater than 0.04 Wm^{-1}K^{-1}.

▼ **Table 6B** Cable surrounded by thermal insulation   Table 52.2

Length in insulation (mm)	Derating factor
50	0.88
100	0.78
200	0.63
400	0.51
≥ 500	0.50

6 Appendix

▼ **Table 6C** Rating factors for one circuit or one multicore cable or for a group of circuits, or a group of multicore cables (to be used with the current-carrying capacities of Tables 6D1, 6E1 and 6F)

Table 4C1

Arrangement (cables touching)	Number of circuits or multicore cables											Applicable reference method for current-carrying capacities
	1	2	3	4	5	6	7	8	9	12	16	20	
Bunched in air, on a surface, embedded or enclosed	1.0	0.80	0.70	0.65	0.60	0.57	0.54	0.52	0.50	0.45	0.41	0.38	A to F
Single layer on wall or floor	1.0	0.85	0.79	0.75	0.73	0.72	0.72	0.71	0.70	0.70	0.70	0.70	C
Single layer multicore on a perforated horizontal or vertical cable tray system	1.0	0.88	0.82	0.77	0.75	0.73	0.73	0.72	0.72	0.72	0.72	0.72	E and F
Single layer multicore on a cable ladder system or cleats, etc.	1.0	0.87	0.82	0.80	0.80	0.79	0.79	0.78	0.78	0.78	0.78	0.78	E and F

Appendix 6

Notes to Table 6C:

1 These factors are applicable to uniform groups of cables, equally loaded.
2 Where horizontal clearances between adjacent cables exceed twice their overall diameter, no rating factor need be applied.
3 The same factors are applied to:
   ▲ groups of two or three single-core cables
   ▲ multicore cables.
4 If a group consists of both two- and three-core cables, the total number of cables is taken as the number of circuits, and the corresponding factor is applied to the Tables for two loaded conductors for the two-core cables, and to the Tables for three loaded conductors for the three-core cables.
5 If a group consists of n single-core cables it may either be considered as $n/2$ circuits of two loaded conductors (for single-phase circuits) or $n/3$ circuits of three loaded conductors (for three-phase circuits).
6 The values given have been averaged over the range of conductor sizes and types of installation included in Tables 4D1A to 4J4A of BS7671 (this includes 6D1, 6E1 and 6F of this guide) and the overall accuracy of tabulated values is within 5 per cent.
7 For some installations and for other methods not provided for in the above table, it may be appropriate to use factors calculated for specific cases, see for example Tables 4C4 and 4C5 of BS 7671.
8 When cables having differing conductor operating temperature are grouped together, the current rating is to be based upon the lowest operating temperature of any cable in the group.
9 If, due to known operating conditions, a cable is expected to carry not more than 30 per cent of its *grouped* rating, it may be ignored for the purpose of obtaining the rating factor for the rest of the group. For example, a group of N loaded cables would normally require a group rating factor of C_g applied to the tabulated I_t. However, if M cables in the group carry loads which are not greater than 0.3 $C_g I_t$ amperes the other cables can be sized by using the group rating factor corresponding to (N minus M) cables.

523.5

6 | Appendix

Table 6D1 Single-core 70 °C thermoplastic (PVC) or thermosetting (note 1) insulated cables, non-armoured, with or without sheath (copper conductors)

Table 4D1A

Ambient temperature: 30 °C
Conductor operating temperature: 70 °C

Current-carrying capacity (amperes):

Conductor cross-sectional area	Reference method A (enclosed in conduit in thermally insulating wall, etc.)		Reference method B (enclosed in conduit on a wall or in trunking, etc.)		Reference method C (clipped direct)			Reference method F (in free air or on a perforated cable tray horizontal or vertical)				
								Touching			Spaced by one cable diameter	
	2 cables, single-phase a.c. or d.c.	3 or 4 cables, three-phase a.c.	2 cables, single-phase a.c. or d.c.	3 or 4 cables, three-phase a.c.	2 cables, single-phase a.c. or d.c. flat and touching	3 or 4 cables, three-phase a.c. flat and touching or trefoil		2 cables, single-phase a.c. or d.c. flat	3 cables, three-phase a.c. flat	3 cables three-phase a.c. trefoil	2 cables single-phase a.c. or d.c. or 3 cables three-phase a.c. flat	
											horizontal	vertical
1	2	3	4	5	6	7		8	9	10	11	12
mm²	A	A	A	A	A	A		A	A	A	A	A
1	11	10.5	13.5	12	15.5	14						
1.5	14.5	13.5	17.5	15.5	20	18						
2.5	20	18	24	21	27	25						
4	26	24	32	28	37	33						
6	34	31	41	36	47	43						
10	46	42	57	50	65	59						
16	61	56	76	68	87	79						
25	80	73	101	89	114	104		131	114	110	146	130
35	99	89	125	110	141	129		162	143	137	181	162
50	119	108	151	134	182	167		196	174	167	219	197
70	151	136	192	171	234	214		251	225	216	281	254
95	182	164	232	207	284	261		304	275	264	341	311

Appendix 6

Table 4E1A

Notes to Table 6D1:
1. The ratings for cables with thermosetting insulation are applicable for cables connected to equipment or accessories designed to operate with cables which run at a temperature not exceeding 70 °C. Where conductor operating temperatures up to 90 °C are acceptable the current rating is increased – see Table 4E1A of BS 7671.
2. Where the conductor is to be protected by a semi-enclosed fuse to BS 3036, see the introduction to this appendix.
3. The current-carrying capacities in columns 2 to 5 are also applicable to flexible cables to BS 6004 Table 1(c) and to 90 °C heat-resisting PVC cables to BS 6231 Tables 8 and 9 where the cables are used in fixed installations.

6 | Appendix

▶ **Table 6D2** Voltage drop (per ampere per metre) at a conductor operating temperature of 70 °C

Table 4D1B

Conductor cross-sectional area	2 cables d.c.	2 cables, single-phase a.c.			3 or 4 cables, three-phase a.c.			
		Reference methods A & B (enclosed in conduit or trunking)	Reference methods C & F (clipped direct on tray or in free air) touching	Reference methods C & F (clipped direct on tray or in free air) spaced	Reference methods A & B (enclosed in conduit or trunking)	Reference methods C & F (clipped direct, on tray or in free air) Touching, Trefoil	Reference methods C & F (clipped direct, on tray or in free air) Touching, Flat	Reference methods C & F (clipped direct, on tray or in free air) Spaced*, Flat
mm²	mV/A/m	mV/A/m	mV/A/m	mV/A/m	mV/A/m	mV/A/m	mV/A/m	mV/A/m
1	44	44	44	44	38	38	38	38
1.5	29	29	29	29	25	25	25	25
2.5	18	18	18	18	15	15	15	15
4	11	11	11	11	9.5	9.5	9.5	9.5
6	7.3	7.3	7.3	7.3	6.4	6.4	6.4	6.4
10	4.4	4.4	4.4	4.4	3.8	3.8	3.8	3.8
16	2.8	2.8	2.8	2.8	2.4	2.4	2.4	2.4
25	1.75	1.80	1.75	1.80	1.55	1.50	1.55	1.55
35	1.25	1.30	1.25	1.30	1.10	1.10	1.10	1.15
50	0.93	1.00	0.95	0.97	0.85	0.82	0.84	0.86
70	0.63	0.72	0.66	0.69	0.61	0.57	0.60	0.63
95	0.46	0.56	0.50	0.54	0.48	0.43	0.47	0.51

* Spacings larger than one cable diameter will result in larger voltage drop.
† The impedance values in Table 6D2 consist of both the resistive and reactive elements of voltage drop usually provided for 25 mm² and above conductor size. For a fuller treatment see Appendix 4 of BS 7671.

Appendix 6

▼ **Table 6E1** Multicore cables having thermoplastic (PVC) or thermosetting insulation (note 1), non-armoured (copper conductors)

Table 4D2A

Ambient temperature: 30 °C
Conductor operating temperature: 70 °C

Current-carrying capacity (amperes):

Conductor cross-sectional area	Reference method A (enclosed in conduit in an insulated wall, etc.)		Reference method B (enclosed in conduit on a wall or in trunking, etc.)		Reference method C (clipped direct)		Reference method E (in free air or on a perforated cable tray, etc. horizontal or vertical)	
	1 two-core cable*, single-phase a.c. or d.c.	1 three-core cable* or 1 four-core cable, three-phase a.c.	1 two-core cable*, single-phase a.c. or d.c.	1 three-core cable* or 1 four-core cable, three-phase a.c.	1 two-core cable*, single-phase a.c. or d.c.	1 three-core cable* or 1 four-core cable, three-phase a.c.	1 two-core cable*, single-phase a.c. or d.c.	1 three-core cable* or 1 four-core cable, three-phase a.c.
1	2	3	4	5	6	7	8	9
mm²	A	A	A	A	A	A	A	A
1	11	10	13	11.5	15	13.5	17	14.5
1.5	14	13	16.5	15	19.5	17.5	22	18.5
2.5	18.5	17.5	23	20	27	24	30	25
4	25	23	30	27	36	32	40	34
6	32	29	38	34	46	41	51	43
10	43	39	52	46	63	57	70	60
16	57	52	69	62	85	76	94	80
25	75	68	90	80	112	96	119	101
35	92	83	111	99	138	119	148	126
50	110	99	133	118	168	144	180	153
70	139	125	168	149	213	184	232	196
95	167	150	201	179	258	223	282	238

6 | Appendix

Notes to Table 6E1:
1 The ratings for cables with thermosetting insulation are applicable for cables connected to equipment or accessories designed to operate with cables which run at a temperature not exceeding 70 °C. Where conductor operating temperatures up to 90 °C are acceptable the current rating is increased – see Table 4E2A of BS 7671. *Table 4E2A*
   Where the conductor is to be protected by a semi-enclosed fuse to BS 3036, to this appendix.
2 * With or without protective conductor. Circular conductors are assumed for sizes up to and including 16 mm². Values for larger sizes relate to shaped conductors and may safely be applied to circular conductors.

Appendix 6

▼ **Table 6E2** Voltage drop (per ampere per metre) at a conductor operating temperature of 70 °C   Table 4D2B

Conductor cross-sectional area	Two-core cable, d.c.	Two-core cable, single-phase a.c.	Three- or four-core cable, three-phase
1	2	3	4
mm²	mV/A/m	mV/A/m	mV/A/m
1	44	44	38
1.5	29	29	25
2.5	18	18	15
4	11	11	9.5
6	7.3	7.3	6.4
10	4.4	4.4	3.8
16	2.8	2.8	2.4
		$z^†$	$z^†$
25	1.75	1.75	1.50
35	1.25	1.25	1.10
50	0.93	0.94	0.81
70	0.63	0.65	0.57
95	0.46	0.50	0.43

† The impedance values in Table 6E2 consist of both the resistive and reactive elements of voltage drop, usually provided separately for 25 mm² and above conductor size. For a fuller treatment see Appendix 4 of BS 7671.

6 | Appendix

▼ **Table 6F** 70 °C thermoplastic (PVC) insulated and sheathed flat cable with protective conductor (copper conductors)

Table 4D5
Ambient temperature: 30 °C
Conductor operating temperature: 70 °C

Current-carrying capacity (amperes) and voltage drop (per ampere per metre):

Conductor cross-sectional area	Reference method 100* (above a plasterboard ceiling covered by thermal insulation not exceeding 100 mm in thickness)	Reference method 101* (above a plasterboard ceiling covered by thermal insulation exceeding 100 mm in thickness)	Reference method 102* (in a stud wall with thermal insulation with cable touching the inner wall surface)	Reference method 103 (in a stud wall with thermal insulation with cable not touching the inner wall surface)	Reference method C (clipped direct)	Reference method A (enclosed in conduit in an insulated wall)	Voltage drop
1	2	3	4	5	6	7	8
mm²	A	A	A	A	A	A	mV/A/m
1	13	10.5	13	8	16	11.5	44
1.5	16	13	16	10	20	14.5	29
2.5	21	17	21	13.5	27	20	18
4	27	22	27	17.5	37	26	11
6	34	27	35	23.5	47	32	7.3
10	45	36	47	32	64	44	4.4
16	57	46	63	42.5	85	57	2.8

Notes:
* Reference methods 100, 101 and 102 require the cable to be in contact with the plasterboard ceiling, wall or joist, see Table 7.2 in Section 7.
1 Wherever practicable, a cable is to be fixed in a position such that it will not be covered with thermal insulation.
2 Regulation 523.7, BS 5803-5: Appendix C: Avoidance of overheating of electric cables, Building Regulations Approved Document B and Thermal insulation: avoiding risks, BR 262, BRE, 2001 refer.

Appendix 7
Certification and reporting

The certificates are used with the kind permission of the BSI.

Introduction

The introduction to Appendix 6 'Model forms for certification and reporting' of BS 7671:2008 is reproduced below.

Appx 6

i   The Electrical Installation Certificate required by Part 6 of BS 7671 should be made out and signed or otherwise authenticated by a competent person or persons in respect of the design, construction, inspection and testing of the work.
ii The Minor Works Certificate required by Part 6 of BS 7671 should be made out and signed or otherwise authenticated by a competent person in respect of the design, construction, inspection and testing of the minor work.
iii The Periodic Inspection Report required by Part 6 of BS 7671 should be made out and signed or otherwise authenticated by a competent person in respect of the inspection and testing of an installation.
iv Competent persons will, as appropriate to their function under **i**, **ii** and **iii** above, have a sound knowledge and experience relevant to the nature of the work undertaken and to the technical standards set down in these Regulations, be fully versed in the inspection and testing procedures contained in these Regulations and employ adequate testing equipment.
v   Electrical Installation Certificates will indicate the responsibility for design, construction, inspection and testing, whether in relation to new work or further work on an existing installation.

Where design, construction, inspection and testing are the responsibility of one person a Certificate with a single signature declaration in the form shown below may replace the multiple signatures section of the model form.

FOR DESIGN, CONSTRUCTION, INSPECTION & TESTING.
I being the person responsible for the Design, Construction, Inspection & Testing of the electrical installation (as indicated by my signature below), particulars of which are described above, having exercised reasonable skill and care when carrying out the Design, Construction, Inspection & Testing, hereby CERTIFY that

7 Appendix

the said work for which I have been responsible is to the best of my knowledge and belief in accordance with BS 7671:2008, amended to ....(date) except for the departures, if any, detailed as follows.

vi A Minor Works Certificate will indicate the responsibility for design, construction, inspection and testing of the work described on the certificate.

vii A Periodic Inspection Report will indicate the responsibility for the inspection and testing of an installation within the extent and limitations specified on the report.

viii A schedule of inspections and a schedule of test results as required by Part 6 should be issued with the associated Electrical Installation Certificate or Periodic Inspection Report.

ix When making out and signing a form on behalf of a company or other business entity, individuals should state for whom they are acting.

x Additional forms may be required as clarification, if needed by ordinary persons, or in expansion, for larger or more complex installations.

xi The IEE Guidance Note 3 provides further information on inspection and testing on completion and for periodic inspections.

Electrical Installation Certificates

Notes for short form (F1) and standard form (F2)

1 The Electrical Installation Certificate is to be used only for the initial certification of a new installation or for an addition or alteration to an existing installation where new circuits have been introduced.

It is not to be used for a Periodic Inspection, for which a Periodic Inspection Report form should be used. For an addition or alteration which does not extend to the introduction of new circuits, a Minor Electrical Installation Works Certificate may be used.

The original Certificate is to be given to the person ordering the work (Regulation 632.3). A duplicate should be retained by the contractor.

2 This Certificate is only valid if accompanied by the Schedule of Inspections and the Schedule(s) of Test Results.

3 The signatures appended are those of the persons authorized by the companies executing the work of design, construction, inspection and testing respectively. A signatory authorized to certify more than one category of work should sign in each of the appropriate places.

4 The time interval recommended before the first periodic inspection must be inserted (see IEE Guidance Note 3 for guidance).

5 The page numbers for each of the Schedules of Test Results should be indicated, together with the total number of sheets involved.

6 The maximum prospective fault current recorded should be the greater of either the short-circuit current or the earth fault current.

7 The proposed date for the next inspection should take into consideration the frequency and quality of maintenance that the installation can reasonably be expected to receive during its intended life, and the period should be agreed between the designer, installer and other relevant parties.

Appendix 7

Form 1   Form No 123 /1

ELECTRICAL INSTALLATION CERTIFICATE (Notes 1 and 2)
(REQUIREMENTS FOR ELECTRICAL INSTALLATIONS - BS 7671 [IEE WIRING REGULATIONS])

DETAILS OF THE CLIENT (note 1)
House Builder Ltd
1 City Road
London

INSTALLATION ADDRESS
Plot 1
New Road
New Town   Postcode AB1 2CD

DESCRIPTION AND EXTENT OF THE INSTALLATION Tick boxes as appropriate

Description of installation: Domestic

New installation ☑

Extent of installation covered by this Certificate:
Complete electrical

Addition to an existing installation ☐

Alteration to an existing installation ☐

...(use continuation sheet if necessary) see continuation sheet No: ...

FOR DESIGN, CONSTRUCTION, INSPECTION & TESTING
I being the person responsible for the Design, Construction, Inspection & Testing of the electrical installation (as indicated by my signature below), particulars of which are described above, having exercised reasonable skill and care when carrying out the Design, Construction, Inspection & Testing, hereby CERTIFY that the said work for which I have been responsible is to the best of my knowledge and belief in accordance with BS 7671:2008 amended to .... (date) except for the departures, if any, detailed as follows:

Details of departures from BS 7671 (Regulations 120.3 and 120.4):
None

The extent of liability of the signatory is limited to the work described above as the subject of this Certificate.

Name (IN BLOCK LETTERS): A. SMITH   Position: Director
Signature (note 3): A. Smith   Date: 01/07/08
For and on behalf of: Electrics Ltd
Address: 27 Central Road
New Town
Postcode EB4 5GH   Tel No: ...

NEXT INSPECTION
I recommend that this installation is further inspected and tested after an interval of not more than ...10... years/~~months~~ (notes 4 and 7)

SUPPLY CHARACTERISTICS AND EARTHING ARRANGEMENTS Tick boxes and enter details, as appropriate

Earthing arrangements		Number and Type of Live Conductors			Nature of Supply Parameters		Supply Protective Device Characteristics	
TN-C	☐	a.c. ☑	d.c. ☐		Nominal voltage, $U/U_0^{(1)}$ 230	V		
TN-S	☐						Type: BS 1361	
TN-C-S	☑	1-phase, 2-wire ☑	2-pole ☐		Nominal frequency, $f^{(1)}$ 50	Hz		
TT	☐	1-phase, 3-wire ☐	3-pole ☐		Prospective fault current, $I_{pf}^{(2)}$ 16	kA	Rated current 100	A
IT	☐				(note 5)			
		2-phase, 3-wire ☐	other ☐		External loop impedance, $Z_e^{(2)}$ 0.35	Ω		
Alternative source of supply (to be detailed on attached schedules) ☐		3-phase, 3-wire ☐			*(Note: (1) by enquiry, (2) by enquiry or by measurement)*			
		3-phase, 4-wire ☐						

Page 1 of 4

7 Appendix

PARTICULARS OF INSTALLATION REFERRED TO IN THE CERTIFICATE	Tick boxes and enter details, as appropriate
Means of Earthing Distributor's facility ☑	**Maximum Demand** Delete as appropriate Maximum demand (load) ......60...... . kVA / Amps

	Details of Installation Earth Electrode *(where applicable)*	
Installation earth electrode ☐	Type (e.g. rod(s), tape etc) ......None......	Location	Electrode resistance to earth .... Ω

Main Protective Conductors

Earthing conductor: material ......copper...... csa ......16......mm² connection verified ☑

Main protective bonding conductors material ......copper...... csa ......10......mm² connection verified ☑

To incoming water and/or gas service ☑ To other elements ....

Main Switch or Circuit-breaker

BS, Type ......BS EN 60947-3...... No. of poles .... Current rating ...80...A Voltage rating ......230......V

Location ......Garage...................................... Fuse rating or setting ....—....A

Rated residual operating current $I_{\Delta n}$ = ..N/A.. mA, and operating time of N/A ms (at $I_{\Delta n}$) *(Applicable only where an RCD is suitable and is used as a main circuit-breaker.)*

COMMENTS ON EXISTING INSTALLATION: *(In the case of an alteration or additions see Section 633)*

.... New installation ....
....
....
....
....
....
....
....
....

SCHEDULES (note 2)
The attached Schedules are part of this document and this Certificate is valid only when they are attached to it.
.....1..... Schedules of Inspections and ....1..... Schedules of Test Results are attached.
(Enter quantities of schedules attached)

GUIDANCE FOR RECIPIENTS

This safety Certificate has been issued to confirm that the electrical installation work to which it relates has been designed, constructed, inspected and tested in accordance with British Standard 7671 (The IEE Wiring Regulations).

You should have received an original Certificate and the contractor should have retained a duplicate Certificate. If you were the person ordering the work, but not the owner of the installation, you should pass this Certificate, or a full copy of it including the schedules, immediately to the owner.

The "original" Certificate should be retained in a safe place and be shown to any person inspecting or undertaking further work on the electrical installation in the future. If you later vacate the property, this Certificate will demonstrate to the new owner that the electrical installation complied with the requirements of British Standard 7671 at the time the Certificate was issued. The Construction (Design and Management) Regulations require that for a project covered by those Regulations, a copy of this Certificate, together with schedules is included in the project health and safety documentation.

For safety reasons, the electrical installation will need to be inspected at appropriate intervals by a competent person. The maximum time interval recommended before the next inspection is stated on Page 1 under "Next Inspection".

This Certificate is intended to be issued only for a new electrical installation or for new work associated with an addition or alteration to an existing installation. It should not have been issued for the inspection of an existing electrical installation. A "Periodic Inspection Report" should be issued for such an inspection.

The Certificate is only valid if a Schedule of Inspections and Schedule of Test Results are appended. Page 2 of 4

Appendix 7

Form 3
Form No 123 /3

SCHEDULE OF INSPECTIONS

Methods of protection against electric shock

Both basic and fault protection:
- [N/A] (i) SELV (Note 1)
- [N/A] (ii) PELV
- [N/A] (iii) Double insulation (Note 2)
- [N/A] (iv) Reinforced insulation (Note 2)

Basic protection: (Note 3)
- [✓] (i) Insulation of live parts
- [✓] (ii) Barriers or enclosures
- [N/A] (iii) Obstacles (Note 4)
- [N/A] (iv) Placing out of reach (Note 5)

Fault protection:
(i) Automatic disconnection of supply:
- [✓] Presence of earthing conductor
- [✓] Presence of circuit protective conductors
- [✓] Presence of protective bonding conductors
- [N/A] Presence of supplementary bonding conductors
- [N/A] Presence of earthing arrangements for combined protective and functional purposes
- [N/A] Presence of adequate arrangements for alternative source(s), where applicable
- [N/A] FELV
- [✓] Choice and setting of protective and monitoring devices (for fault and/or overcurrent protection)

(ii) Non-conducting location: (Note 6)
- [N/A] Absence of protective conductors

(iii) Earth-free local equipotential bonding: (Note 6)
- [N/A] Presence of earth-free local equipotential bonding

(iv) Electrical Separation: (Note 7)
- [N/A] Provided for **one item** of current-using equipment
- [N/A] Provided for **more than one item** of current-using equipment

Additional protection:
- [✓] Presence of residual current devices(s)
- [N/A] Presence of supplementary bonding conductors

Prevention of mutual detrimental influence
- [✓] (a) Proximity of non-electrical services and other influences
- [✓] (b) Segregation of Band I and Band II circuits or use of Band II insulation
- [✓] (c) Segregation of safety circuits

Identification
- [✓] (a) Presence of diagrams, instructions, circuit charts and similar information
- [✓] (b) Presence of danger notices and other warning notices
- [✓] (c) Labelling of protective devices, switches and terminals
- [✓] (d) Identification of conductors

Cables and conductors
- [✓] Selection of conductors for current-carrying capacity and voltage drop
- [✓] Erection methods
- [✓] Routing of cables in prescribed zones
- [✓] Cables incorporating earthed armour or sheath, or run within an earthed wiring system, or otherwise adequately protected against nails, screws and the like
- [✓] Additional protection provided by 30 mA RCD for cables in concealed walls (where required in premises not under the supervision of a skilled or instructed person)
- [✓] Connection of conductors
- [✓] Presence of fire barriers, suitable seals and protection against thermal effects

General
- [✓] Presence and correct location of appropriate devices for isolation and switching
- [✓] Adequacy of access to switchgear and other equipment
- [✓] Particular protective measures for special installations and locations
- [✓] Connection of single-pole devices for protection or switching in line conductors only
- [✓] Correct connection of accessories and equipment
- [N/A] Presence of undervoltage protective devices
- [✓] Selection of equipment and protective measures appropriate to external influences
- [✓] Selection of appropriate functional switching devices

Inspected by ...A. Smith...... Date ...01/07/08......

Notes:
- ✓ to indicate an inspection has been carried out and the result is satisfactory
- X to indicate an inspection has been carried out and the result is not satisfactory (applicable to a periodic inspection only)
- N/A to indicate the inspection is not applicable to a particular item
- LIM to indicate that, exceptionally, a limitation agreed with the person ordering the work prevented the inspection or test being carried out (applicable to a periodic inspection only).

1. SELV – an extra-low voltage system which is electrically separated from Earth and from other systems in such a way that a single-fault cannot give rise to the risk of electric shock. The particular requirements of the Regulations must be checked (see Section 414)
2. Double or reinforced insulation. Not suitable for domestic or similar installations if it is the sole protective measure (see 412.1.3).
3. Basic protection – will include measurement of distances where appropriate
4. Obstacles – only adopted in special circumstances (see 417.2)
5. Placing out of reach – only adopted in special circumstances (see 417.3)
6. Non-conducting locations and Earth-free local equipotential bonding – these are not recognised for general application. May only be used where the installation is controlled/under the supervision of skilled or instructed persons (see Section 418)
7. Electrical separation – the particular requirements of the Regulations must be checked. If a single item of current-using equipment is supplied from a single source, see Section 413. If more than one item of current-using equipment is supplied from a single source then the installation must be controlled/under the supervision of skilled or instructed persons, see also Regulation 418.3

7 | Appendix

Form 4
SCHEDULE OF TEST RESULTS

Form No 123 /4

Contractor: Electrics Ltd
Test Date: 01/07/08
Signature: A. Smith
Method of fault protection: automatic disconnection of supply
Equipment vulnerable to testing: Two RCDs, shower, luminaires and dimmer in lounge

Address/Location of distribution board: Plot 1, New Road, New Town

*1 Type of Supply: TN-S/TN-C-S/TT
*2 Ze at origin: 0.35 ohms
*3 PFC: 16 kA
Confirmation of supply polarity ☑

Instruments
loop impedance: AB 11
continuity: AB 22
insulation: AB 44
RCD tester: AB 55

Description of Work: Dwelling elec. installation

Circuit Description	Overcurrent Device			Wiring Conductors			Continuity				Insulation Resistance		P o l a r i t y	Test Results			Remarks
	type	Rating I_n	*4 Short-circuit capacity ...6... kA	live	cpc		$(R_1 + R_2)$*	R_2*	R i n g *8		Live/Live	Live/Earth		Earth Loop Impedance Z_s	Functional Testing		
		A		mm²	mm²		Ω *6	Ω *7			MΩ *9	MΩ *10	*11	Ω *12	RCD time ms *13	Other *14	
1	2	3		4	5												15
RCD 1		30mA															
Lights up	B	10		1.5	1.0		2.4	—			40	30	✓	2.7		✓	
Sockets down	B	32		2.5	1.5		0.4	—	✓		50	40	✓	0.7		✓	
Cooker	B	32		6.0	2.5		0.3	—			50	40	✓	0.6		✓	
RCD 2		30mA													26		
Lights down	B	10		1.5	1.0		2.2	—			—	30	✓	2.5		✓	dimmer, luminaire
Sockets up	B	32		2.5	1.5		0.6	—	✓		30	30	✓	0.9		✓	electric shower
Shower	B	40		10	4.0		0.2	—			—	40	✓	0.5	24	✓	

Deviations from Wiring Regulations and special notes:

None

*Number - See notes on schedule of test results on page 141 of Guide **Complete column 6 or 7**

Page 4 of 4

Appendix **7**

Schedule of Test Results

Notes

*1 Type of supply is ascertained from the distributor or by inspection.

*2 Z_e at origin. When the maximum value declared by the distributor is used, the effectiveness of the earth must be confirmed by a test. If measured the main bonding will need to be disconnected for the duration of the test.

*3 Prospective fault current (PFC). The value recorded is the greater of either the short-circuit current or the earth fault current. Preferably determined by enquiry of the distributor.

*4 Short-circuit capacity of the device is noted, see Table 7.4 of the *On-Site Guide* or Table 2.4 of GN3.

The following tests, where relevant, must be carried out in the following sequence:

*5 Continuity of protective conductors, including main and supplementary bonding
Every protective conductor, including main and supplementary bonding conductors, should be tested to verify that it is continuous and correctly connected.

*6 Continuity
Where test method 1 is used, enter the measured resistance of the line conductor plus the circuit protective conductor ($R_1 + R_2$). See 10.3.1 of the *On-Site Guide* or 2.7.5 of GN3. During the continuity testing (test method 1) the following polarity checks should be carried out:
 a every fuse and single-pole control and protective device is connected in the line conductor only
 b centre-contact bayonet and Edison screw lampholders have outer contact connected to the neutral conductor
 c wiring is correctly connected to socket-outlets and similar accessories.
Compliance is indicated by a tick in polarity column 11.
($R_1 + R_2$) need not be recorded if R_2 is recorded in column 7.

*7 Where test method 2 is used, the maximum value of R2 is recorded in column 7. See 10.3.1 of the *On-Site Guide* or 2.7.5 of GN3.

*8 Continuity of ring final circuit conductors
A test must be made to verify the continuity of each conductor including the protective conductor of every ring final circuit. See 10.3.2 of the *On-Site Guide* or 2.7.6 of GN3.

*9, *10 Insulation resistance
All voltage sensitive devices to be disconnected or test between live conductors (line and neutral) connected together and earth. The insulation resistance between live conductors is inserted in column 9 and between live conductors and earth in column 10.

The minimum insulation resistance values are given in Table 10.1 of the *On-Site Guide* or Table 2.2 of GN3. See 10.3.3 of the *On-Site Guide* or 2.7.7 of GN3.

7 Appendix

All the preceding tests should be carried out before the installation is energised.

*11 Polarity
A satisfactory polarity test may be indicated by a tick in column 11. Only in a Schedule of Test Results associated with a Periodic Inspection Report is it acceptable to record incorrect polarity.

Note: Correct polarity of the supply should be confirmed and indicated by a tick in the box at the top of the schedule.

*12 Earth fault loop impedance Z_s
This may be determined either by direct measurement at the furthest point of a live circuit or by adding $(R_1 + R_2)$ of column 6 to Z_e. Z_e is determined by measurement at the origin of the installation or preferably the value declared by the supply company used. $Z_s = Z_e + (R_1 + R_2)$. Z_s should not exceed the values given in Appendix 2 of the *On-Site Guide* or Appendix B of GN3.

*13 Functional testing
The operation of RCDs (including RCBOs) is tested by simulating a fault condition, independent of any test facility in the device. Record operating time in column 13. Effectiveness of the test button must be confirmed. See Section 11 of the *On-Site Guide* or 2.7.15 and 2.7.18 of GN3.

*14 All switchgear and controlgear assemblies, drives, control and interlocks, etc. must be operated to ensure that they are properly mounted, adjusted and installed. Satisfactory operation is indicated by a tick in column 14.

Earth electrode resistance
The earth electrode resistance of TT installations must be measured, and normally an RCD is required. For reliability in service the resistance of any earth electrode should be below 200 Ω. Record the value on Form 1, 2 or 6, as appropriate. See 10.3.5 of the *On-Site Guide* or 2.7.12 of GN3.

Appendix 7

Form 2
Form No 124 /2

ELECTRICAL INSTALLATION CERTIFICATE (notes 1 and 2)
(REQUIREMENTS FOR ELECTRICAL INSTALLATIONS - BS 7671 [IEE WIRING REGULATIONS])

DETAILS OF THE CLIENT (note 1)	A Developer Ltd High Street Town, County

INSTALLATION ADDRESS	Plot 10, Industrial Estate Town County .... Postcode ....

DESCRIPTION AND EXTENT OF THE INSTALLATION Tick boxes as appropriate
(note 1)
Description of installation: **Industrial with office**

New installation ☑

Extent of installation covered by this Certificate: **Complete installation**

Addition to an existing installation ☐

Alteration to an existing installation ☐

.... (use continuation sheet if necessary) see continuation sheet No: ......

FOR DESIGN
I/We being the person(s) responsible for the design of the electrical installation (as indicated by my/our signatures below), particulars of which are described above, having exercised reasonable skill and care when carrying out the design hereby CERTIFY that the design work for which I/we have been responsible is to the best of my/our knowledge and belief in accordance with BS 7671:2008, amended to..........................(date) except for the departures, if any, detailed as follows:

Details of departures from BS 7671 (Regulations 120.3 and 120.4):
None

The extent of liability of the signatory or the signatories is limited to the work described above as the subject of this Certificate.

For the DESIGN of the installation: **(Where there is mutual responsibility for the design)

Signature: *B. Brown* Date: **08/07/08** Name (BLOCK LETTERS): **B. BROWN** Designer No 1

Signature: — Date: — Name (BLOCK LETTERS): — Designer No 2**

FOR CONSTRUCTION
I/We being the person(s) responsible for the construction of the electrical installation (as indicated by my/our signatures below), particulars of which are described above, having exercised reasonable skill and care when carrying out the construction hereby CERTIFY that the construction work for which I/we have been responsible is to the best of my/our knowledge and belief in accordance with BS 7671:2008, amended to ....(date) except for the departures, if any, detailed as follows:

Details of departures from BS 7671 (Regulations 120.3 and 120.4):
None

The extent of liability of the signatory is limited to the work described above as the subject of this Certificate.

For CONSTRUCTION of the installation:
Signature: *W. White* Date: **09/07/08**
Name (BLOCK LETTERS): **W. WHITE** Constructor

FOR INSPECTION & TESTING
I/We being the person(s) responsible for the inspection & testing of the electrical installation (as indicated by my/our signatures below), particulars of which are described above, having exercised reasonable skill and care when carrying out the inspection & testing hereby CERTIFY that the work for which I/we have been responsible is to the best of my/our knowledge and belief in accordance with BS 7671:2008, amended to ....(date) except for the departures, if any, detailed as follows:

Details of departures from BS 7671 (Regulations 120.3 and 120.4):
None

The extent of liability of the signatory is limited to the work described above as the subject of this Certificate.

For INSPECTION AND TEST of the installation:
Signature: *S. Jones* Date: **11/07/08**
Name (BLOCK LETTERS): **S. JONES** Inspector

NEXT INSPECTION (notes 4 and 7)
I/We the designer(s), recommend that this installation is further inspected and tested after an interval of not more than **3** years/~~months~~.

Page 1 of 4

7 Appendix

PARTICULARS OF SIGNATORIES TO THE ELECTRICAL INSTALLATION CERTIFICATE (note 3)

Designer (No 1)
Name: B. Brown
Address: City Road, Old Town, County
Company: Design Co. Ltd
Postcode: XY10 2ZX   Tel No: 01334 261226

Designer (No 2) (if applicable)
Name: Not applicable
Address:
Company:
Postcode: Tel No:

Constructor
Name: W. White
Address: 187 Industrial Lane, Town
Company: County Electrics Ltd
Postcode: XYZ 4AB   Tel No: 01334 261337

Inspector
Name: S. Jones
Address: As above
Company: County Electrics Ltd
Postcode: Tel No:

SUPPLY CHARACTERISTICS AND EARTHING ARRANGEMENTS
Tick boxes and enter details, as appropriate

Earthing arrangements
- TN-C ☐
- TN-S ☐
- TN-C-S ☑
- TT ☐
- IT ☐

Alternative source ☐ of supply (to be detailed on attached schedules)

Number and Type of Live Conductors
- a.c. ☐   d.c. ☐
- 1-phase, 2-wire ☐   2-pole ☐
- 1-phase, 3-wire ☐   3-pole ☐
- 2-phase, 3-wire ☐   other ☐
- 3-phase, 3-wire ☐
- 3-phase, 4-wire ☑

Nature of Supply Parameters
- Nominal voltage, U/U_0 [1] 400 / 230 V
- Nominal frequency, f [1] 50 Hz
- Prospective fault current, I_{pf} [2] 18 kA (note 6)
- External loop impedance, Z_e [2] 0·2 Ω

(Note: (1) by enquiry, (2) by enquiry or by measurement)

Supply Protective Device Characteristics
- Type: BS 1361 Type II
- Rated current 100 A

PARTICULARS OF INSTALLATION REFERRED TO IN THE CERTIFICATE
Tick boxes and enter details, as appropriate

Means of Earthing
Distributor's facility ☑

Maximum Demand
Maximum demand (load) 40 A / phase   ~~kVA~~/ Amps *Delete as appropriate*

Details of Installation Earth Electrode (where applicable)

Installation earth electrode ☐	Type (e.g. rod(s), tape etc)	Location	Electrode resistance to Earth
	—	—	— Ω

Main Protective Conductors

Earthing conductor: material copper   csa 16 mm²   connection verified ☑

Main protective bonding conductors   material copper   csa 10 mm²   connection verified ☑
To incoming water and/or gas service ☑   To other elements

Main Switch or Circuit-breaker
BS, Type BS EN 60947-3   No. of poles 3   Current rating 125 A   Voltage rating 400 V
Location Switchroom   Fuse rating or setting — A
Rated residual operating current $I_{\Delta n}$ = ... mA, and operating time of ...... ms (at $I_{\Delta n}$) *(applicable only where an RCD is suitable and is used as a main circuit-breaker)*

COMMENTS ON EXISTING INSTALLATION: *(In the case of an alteration or additions see Section 633)*

Not applicable

SCHEDULES (note 2)
The attached Schedules are part of this document and this Certificate is valid only when they are attached to it.
1 Schedules of Inspections and 1 Schedules of Test Results are attached.
(Enter quantities of schedules attached.)

Page 2 of 4 (note 5)

Appendix 7

Electrical Installation Certificate

Guidance for recipients (to be appended to the Certificate)

This safety Certificate has been issued to confirm that the electrical installation work to which it relates has been designed, constructed, inspected and tested in accordance with British Standard 7671 (The IEE Wiring Regulations).

You should have received an original Certificate and the contractor should have retained a duplicate Certificate. If you were the person ordering the work, but not the owner of the installation, you should pass this Certificate, or a full copy of it including the schedules, immediately to the owner.

The 'original' Certificate should be retained in a safe place and be shown to any person inspecting or undertaking further work on the electrical installation in the future. If you later vacate the property, this Certificate will demonstrate to the new owner that the electrical installation complied with the requirements of British Standard 7671 at the time the Certificate was issued. The Construction (Design and Management) Regulations require that, for a project covered by those Regulations, a copy of this Certificate, together with schedules, is included in the project health and safety documentation.

For safety reasons, the electrical installation will need to be inspected at appropriate intervals by a competent person. The maximum time interval recommended before the next inspection is stated on Page 1 under 'Next Inspection'.

This Certificate is intended to be issued only for a new electrical installation or for new work associated with an addition or alteration to an existing installation. It should not have been issued for the inspection of an existing electrical installation. A Periodic Inspection Report should be issued for such an inspection.

7 Appendix

Form 3
Form No 124 /3

SCHEDULE OF INSPECTIONS

Methods of protection against electric shock

Both basic and fault protection:
- [N/A] (i) SELV (Note 1)
- [N/A] (ii) PELV
- [N/A] (iii) Double insulation (Note 2)
- [N/A] (iv) Reinforced insulation (Note 2)

Basic protection: (Note 3)
- [✓] (i) Insulation of live parts
- [✓] (ii) Barriers or enclosures
- [N/A] (iii) Obstacles (Note 4)
- [✓] (iv) Placing out of reach (Note 5)

Fault protection:
(i) Automatic disconnection of supply:
- [✓] Presence of earthing conductor
- [✓] Presence of circuit protective conductors
- [✓] Presence of protective bonding conductors
- [N/A] Presence of supplementary bonding conductors
- [✓] Presence of earthing arrangements for combined protective and functional purposes
- [✓] Presence of adequate arrangements for alternative source(s), where applicable
- [N/A] FELV
- [✓] Choice and setting of protective and monitoring devices (for fault and/or overcurrent protection)

(ii) Non-conducting location: (Note 6)
- [N/A] Absence of protective conductors

(iii) Earth-free local equipotential bonding: (Note 6)
- [N/A] Presence of earth-free local equipotential bonding

(iv) Electrical Separation: (Note 7)
- [N/A] Provided for **one item** of current-using equipment
- [N/A] Provided for **more than one item** of current-using equipment

Additional protection:
- [N/A] Presence of residual current devices(s)
- [N/A] Presence of supplementary bonding conductors

Prevention of mutual detrimental influence
- [✓] (a) Proximity of non-electrical services and other influences
- [✓] (b) Segregation of Band I and Band II circuits or use of Band II insulation
- [✓] (c) Segregation of safety circuits

Identification
- [✓] (a) Presence of diagrams, instructions, circuit charts and similar information
- [✓] (b) Presence of danger notices and other warning notices
- [✓] (c) Labelling of protective devices, switches and terminals
- [✓] (d) Identification of conductors

Cables and conductors
- [✓] Selection of conductors for current-carrying capacity and voltage drop
- [✓] Erection methods
- [✓] Routing of cables in prescribed zones
- [✓] Cables incorporating earthed armour or sheath, or run within an earthed wiring system, or otherwise adequately protected against nails, screws and the like
- [N/A] Additional protection provided by 30 mA RCD for cables in concealed walls (where required in premises not under the supervision of a skilled or instructed person)
- [✓] Connection of conductors
- [✓] Presence of fire barriers, suitable seals and protection against thermal effects

General
- [✓] Presence and correct location of appropriate devices for isolation and switching
- [✓] Adequacy of access to switchgear and other equipment
- [N/A] Particular protective measures for special installations and locations
- [✓] Connection of single-pole devices for protection or switching in line conductors only
- [✓] Correct connection of accessories and equipment
- [N/A] Presence of undervoltage protective devices
- [✓] Selection of equipment and protective measures appropriate to external influences
- [✓] Selection of appropriate functional switching devices

Inspected by ...... S. Jones ...... Date ...... 11/07/08 ......

Notes:
- ✓ to indicate an inspection has been carried out and the result is satisfactory
- X to indicate an inspection has been carried out and the result is not satisfactory (applicable to a periodic inspection only)
- N/A to indicate the inspection is not applicable to a particular item
- LIM to indicate that, exceptionally, a limitation agreed with the person ordering the work prevented the inspection or test being carried out (applicable to a periodic inspection only).

1. SELV – an extra-low voltage system which is electrically separated from Earth and from other systems in such a way that a single-fault cannot give rise to the risk of electric shock. The particular requirements of the Regulations must be checked (see Section 414)
2. Double or reinforced insulation. Not suitable for domestic or similar installations if it is the sole protective measure (see 412.1.3).
3. Basic protection – will include measurement of distances where appropriate
4. Obstacles – only adopted in special circumstances (see 417.2)
5. Placing out of reach – only adopted in special circumstances (see 417.3)
6. Non-conducting locations and Earth-free local equipotential bonding – these are not recognised for general application. May only be used where the installation is controlled/under the supervision of skilled or instructed persons (see Section 418)
7. Electrical separation – the particular requirements of the Regulations must be checked. If a single item of current-using equipment is supplied from a single source, see Section 413. If more than one item of current-using equipment is supplied from a single source then the installation must be controlled/under the supervision of skilled or instructed persons, see also Regulation 418.3

Page 3 of 4

Appendix 7

Form 4
SCHEDULE OF TEST RESULTS

Contractor: County Electrics Ltd
Test Date: 11/07/08
Signature: S. Jones
Method of fault protection: automatic disconnection of supply
Equipment vulnerable to testing: luminaires, lighting controller

Address/Location of distribution board: Plot 10, Industrial Estate

*1 Type of Supply: ~~TN-S~~/TN-C-S/~~TT~~
*2 Ze at origin: 0.2 ohms
*3 PFC: 18 kA
Confirmation of supply polarity ☑

Form No 124/4

Instruments
loop impedance: AB.11
continuity: AB.27
insulation: AB.44
RCD tester: AB.65

Description of Work: Speculative industrial with office

Circuit Description	Overcurrent Device			Wiring Conductors			Continuity				Insulation Resistance		Polarity	Earth Loop Impedance Z_s	Test Results — RCD time	Functional Testing	Other	Remarks
	type	Rating I_n	*4 Short-circuit capacity: 2.5 kA	live	cpc		(R_1+R_2)*	R_2*	$R_i n g$*		Live/Live	Live/Earth						
		A		mm²	mm²		Ω *6	Ω *7	*8		MΩ *9	MΩ *10	*11	Ω *12	ms *13	*14		*15
1	2	3		4	5													
Lights 1	=	16		2.5	1.5		2.0	—	—		—	10	✓	2.2		✓		vulnerable
Lights 2	=	16		2.5	1.5		2.3	—	—		—	10	✓	2.5		✓		vulnerable
Lights 3	=	16		2.5	1.5		1.6	—	—		—	10	✓	1.8		✓		vulnerable
Sockets 1	=	32		2.5	1.5		0.5	—	✓		30	20	✓	0.7		✓		
Sockets 2	=	32		2.5	1.5		0.4	—	✓		30	20	✓	0.6		✓		
Busbar 1	=	63		16	10		0.1	—	—		40	20	✓	0.3				
Busbar 2	=	63		16	10		0.1	—	—		40	20	✓	0.3				

Deviations from Wiring Regulations and special notes: Special note:
The installation is to be under the supervision of skilled or instructed persons, there is no RCD protection to socket-outlets or switchdrops.

*Number - See notes on schedule of test results on page 141 of Guide   *Complete column 6 or 7

Page 4 of 4

7 Appendix

Minor Electrical Installation Works Certificate
Notes on completion

Scope
The Minor Works Certificate is intended to be used for additions and alterations to an installation that do not extend to the provision of a new circuit. Examples include the addition of socket-outlets or lighting points to an existing circuit, the relocation of a light switch etc. This Certificate may also be used for the replacement of equipment such as accessories or luminaires, but not for the replacement of distribution boards or similar items. Appropriate inspection and testing, however, should always be carried out irrespective of the extent of the work undertaken.

Part 1 Description of minor works
1,2 The minor works must be so described that the work that is the subject of the certification can be readily identified.

4 See Regulations 120.3 and 120.4. No departures are to be expected except in most unusual circumstances.

Part 2 Installation details
2 The method of fault protection must be clearly identified e.g. automatic disconnection of supply (ADS) using fuse, circuit-breaker or RCD.

4 If the existing installation lacks either an effective means of earthing or adequate main protective bonding conductors, this must be clearly stated. See Regulation 633.2.

Recorded departures from BS 7671 may constitute non-compliance with the Electricity Safety, Quality and Continuity Regulations 2002 (as amended) or the Electricity at Work Regulations 1989. It is important that the client is advised immediately in writing.

Part 3 Essential tests
The relevant provisions of Part 6 (Inspection and Testing) of BS 7671 must be applied in full to all minor works. For example, where a socket-outlet is added to an existing circuit it is necessary to:

1. establish that the earthing contact of the socket-outlet is connected to the main earthing terminal
2. measure the insulation resistance of the circuit that has been added to, and establish that it complies with Table 61 of BS 7671
3. measure the earth fault loop impedance to establish that the maximum permitted disconnection time is not exceeded
4. check that the polarity of the socket-outlet is correct
5. (if the work is protected by an RCD) verify the effectiveness of the RCD.

Appendix 7

Part 4 Declaration

1,3 The Certificate must be made out and signed by a competent person in respect of the design, construction, inspection and testing of the work.

1,3 The competent person will have a sound knowledge and experience relevant to the nature of the work undertaken and to the technical standards set down in BS 7671, be fully versed in the inspection and testing procedures contained in the Regulations and employ adequate testing equipment.

2 When making out and signing a form on behalf of a company or other business entity, individuals must state for whom they are acting.

7 | Appendix

Form 5
Form No **125** /5

MINOR ELECTRICAL INSTALLATION WORKS CERTIFICATE
(REQUIREMENTS FOR ELECTRICAL INSTALLATIONS - BS 7671 [IEE WIRING REGULATIONS])
To be used only for minor electrical work which does not include the provision of a new circuit

PART 1 : Description of minor works
1. Description of the minor works : **Additional socket in main office**
2. Location/Address : **Office Co. Ltd, New Town Road**
3. Date minor works completed : **21/06/08**
4. Details of departures, if any, from BS 7671

   None

PART 2 : Installation details
1. System earthing arrangement: TN-C-S ☑ TN-S ☐ TT ☐
2. Method of fault protection: **automatic disconnection of supply by circuit-breaker**
3. Protective device for the modified circuit : Type BS **EN 60947-2** Rating **32** A
4. Comments on existing installation, including adequacy of earthing and bonding arrangements : (See Regulation 131.8)

   Socket circuit not RCD protected as client informs that the installation is under the supervision of a skilled person.

PART 3 : Essential Tests
1. Earth continuity : satisfactory ☑
2. Insulation resistance:
   - Line/neutral ......**> 200**...... MΩ
   - Line/earth ......**> 200**...... MΩ
   - Neutral/earth ......**> 200**...... MΩ
3. Earth fault loop impedance **0·8** Ω
4. Polarity : satisfactory ☑
5. RCD operation (if applicable): Rated residual operating current I$_{\Delta n}$ ....mA and operating time of ....ms (at I$_{\Delta n}$)

PART 4 : Declaration
1. I/We CERTIFY that the said works do not impair the safety of the existing installation, that the said works have been designed, constructed, inspected and tested in accordance with BS 7671:2008 (IEE Wiring Regulations), amended to ....(date) and that the said works, to the best of my/our knowledge and belief, at the time of my/our inspection, complied with BS 7671 except as detailed in Part 1 above.
2. Name: **W. White**   3. Signature: **W. White**
   For and on behalf of: **County Electrics Ltd**   Position: **Electrician**
   Address: **187 Industrial Lane Town**
   Date: **21/06/08**

GUIDANCE FOR RECIPIENTS

This safety Certificate has been issued to confirm that the electrical installation work to which it relates has been designed, constructed, inspected and tested in accordance with British Standard 7671 (The IEE Wiring Regulations).

You should have received an original Certificate and the contractor should have retained a duplicate Certificate. If you were the person ordering the work, but not the owner of the installation, you should pass this Certificate, or a full copy of it, immediately to the owner.

A separate Certificate should have been received for each existing circuit on which minor works have been carried out. This Certificate is not appropriate if you requested the contractor to undertake more extensive installation work, for which you should have received an Electrical Installation Certificate.

The "original" Certificate should be retained in a safe place and be shown to any person inspecting or undertaking further work on the electrical installation in the future. If you later vacate the property, this Certificate will demonstrate to the new owner that the minor electrical installation work carried out complied with the requirements of British Standard 7671 at the time the Certificate was issued.

Appendix 7

Periodic Inspection Report

Notes

1. This Periodic Inspection Report form should only be used for reporting on the condition of an existing installation.
2. The Report, normally comprising at least four pages, should include schedules of both the inspection and the test results. Additional sheets of test results may be necessary for other than a simple installation. The page numbers of each sheet should be indicated, together with the total number of sheets involved.
3. The intended purpose of the Periodic Inspection Report should be identified, together with the recipient's details, in the appropriate boxes.
4. The maximum prospective fault current recorded should be the greater of either the short-circuit current or the earth fault current.
5. The 'Extent and Limitations' box should fully identify the elements of the installation that are covered by the report and those that are not, this aspect having been agreed with the client and other interested parties before the inspection and testing is carried out.
6. The recommendation(s), if any, should be categorised using the numbered coding 1-4 as appropriate.
7. The 'Summary of the Inspection' box should clearly identify the condition of the installation in terms of safety.
8. Where the periodic inspection and testing has resulted in a satisfactory overall assessment, the time interval for the next periodic inspection and testing should be given. The IEE Guidance Note 3 provides guidance on the maximum interval between inspections for various types of buildings. If the inspection and testing reveal that parts of the installation require urgent attention, it would be appropriate to state an earlier re-inspection date, having due regard to the degree of urgency and extent of the necessary remedial work.
9. If the space available on the model form for information on recommendations is insufficient, additional pages should be provided as necessary.

7 Appendix

Form 6   Form No 126 /6

PERIODIC INSPECTION REPORT FOR AN ELECTRICAL INSTALLATION (note 1)
(REQUIREMENTS FOR ELECTRICAL INSTALLATIONS - BS 7671 [IEE WIRING REGULATIONS])

DETAILS OF THE CLIENT
Client: Mr A. Brown
Address: 111 Any Street, Town, County PO11 CO2
Purpose for which this Report is required: Mortgage application .....(note 3)

DETAILS OF THE INSTALLATION Tick boxes as appropriate
Occupier: As above
Installation:
Address:
Description of Premises: Domestic ☑   Commercial ☐   Industrial ☐   Other ☐
House with garage
Estimated age of the Electrical Installation: 25 years
Evidence of Alterations or Additions: Yes ☑   No ☐   Not apparent ☐
If "Yes", estimate age: 3 years
Date of last inspection: —   Records available   Yes ☐   No ☑

EXTENT AND LIMITATIONS OF THE INSPECTION (note 5)
Extent of electrical installation covered by this report: House, garage, garden shed

Limitations: (see Regulation 634.2)
No dismantling or lifting of floorboards

This inspection has been carried out in accordance with BS 7671 : 2008 (IEE Wiring Regulations), amended to .... Cables concealed within trunking and conduits, or cables and conduits concealed under floors, in roof spaces and generally within the fabric of the building or underground have not been inspected.

NEXT INSPECTION (note 8)
I/We recommend that this installation is further inspected and tested after an interval of not more than 5 ~~months~~/years, provided that any observations 'requiring urgent attention' are attended to without delay.

DECLARATION
INSPECTED AND TESTED BY
Name: W. White   Signature: W. White
For and on behalf of: County Electrics Ltd   Position: Electrician
Address: 187 Industrial Estate
         Town   Date: 26/08/08
         County

Page 1 of 4

Appendix 7

SUPPLY CHARACTERISTICS AND EARTHING ARRANGEMENTS
Tick boxes and enter details, as appropriate

Earthing arrangements	Number and Type of Live Conductors	Nature of Supply Parameters	Supply Protective Device Characteristics
TN-C ☐ TN-S ☐ TN-C-S ☑ TT ☐ IT ☐	a.c. ☑ d.c. ☐ 1-phase, 2-wire ☑ 2-pole ☐ 1-phase, 3 wire ☐ 3-pole ☐	Nominal voltage, $U/U_0^{(1)}$ ... 230 ... V Nominal frequency, $f^{(1)}$ ... 50 ... Hz Prospective fault current, $I_{pf}^{(2)}$ ... 1·0 ... kA (note 4) External loop impedance, $Z_e^{(2)}$ ... 0·24 ... Ω	Type: BS 1361 Rated current: ... 100 ... A
Alternative source ☐ of supply (to be detailed on attached schedules)	2-phase, 3-wire ☐ other ☐ 3-phase, 3-wire ☐ 3-phase, 4-wire ☐	*(Note: (1) by enquiry, (2) by enquiry or by measurement)*	

PARTICULARS OF INSTALLATION REFERRED TO IN THE REPORT
Tick boxes and enter details, as appropriate

Means of Earthing
Distributor's facility ☑
Installation earth electrode ☐

Details of Installation Earth Electrode *(where applicable)*
Type (e.g. rod(s), tape etc) ..... —
Location .....
Electrode resistance to Earth ..... — ..... Ω

Main Protective Conductors
Earthing conductor: material ... copper ... csa ... 10 ... mm² connection verified ☑
Main equipotential bonding conductors material ... copper ... csa ... 6 ... mm² connection verified ☑

To incoming water service ☑   To incoming gas service ☑   To incoming oil service ☐   To structural steel ☐
To lightning protection ☐   To other incoming service(s) ☐   (state details .....)

Main Switch or Circuit-breaker
BS, Type ... BS 5486 .....   No. of poles ... 2 ...   Current rating ... 80 ... A   Voltage rating ... 240 ... V
Location ... Meter cupboard .....   Fuse rating or setting ... — ... A
Rated residual operating current $I_{\Delta n}$ = ... — ... mA, and operating time of ... — ... ms (at $I_{\Delta n}$) *(applicable only where an RCD is suitable and is used as a main circuit-breaker)*

OBSERVATIONS AND RECOMMENDATIONS
Tick boxes as appropriate (note 9)

Referring to the attached Schedule(s) of Inspections and Test Results, and subject to the limitations specified at the Extent and Limitations of the Inspection section

☐ No remedial work is required   ☑ The following observations are made:

Recommendations as detailed below: note 6

1)	Broken socket-outlet in kitchen — accessible live parts	1
2)	Lighting pendants and lampholders worn / overheating	1
3)	No RCD to socket-outlets	2
4)	No RCD or supplementary bonding to bathroom circuits	2
5)	Main bonding conductor size 6 mm² in a TN-C-S installation	4
6)	Consumer unit not labelled	4

One of the following numbers, as appropriate, is to be allocated to each of the observations made above to indicate to the person(s) responsible for the installation the action recommended.

[1] requires urgent attention   [2] requires improvement   [3] requires further investigation
[4] does not comply with BS 7671: 2008 amended to ..... This does not imply that the electrical installation inspected is unsafe.

SUMMARY OF THE INSPECTION (note 7)
Date(s) of the inspection: ... 26 / 8 / 2008
General condition of the installation: ..... The broken socket and worn pendant / overheating lampholders require urgent attention. RCD protection is required to the socket-outlet circuits and the bathroom circuits

Overall assessment: ~~Satisfactory~~/Unsatisfactory (note 8)   see above

SCHEDULE(S)
The attached Schedules are part of this document and this Report is valid only when they are attached to it.
..... 1 ..... Schedules of Inspections and ..... 1 ..... Schedules of Test Results are attached.
(Enter quantities of schedules attached).

Page 2 of 4

7 Appendix

Periodic Inspection Report

Guidance for recipients (to be appended to the Report)

This Periodic Inspection Report form is intended for reporting on the condition of an existing electrical installation.

You should have received an original Report and the contractor should have retained a duplicate. If you were the person ordering this Report, but not the owner of the installation, you should pass this Report, or a copy of it, immediately to the owner.

The 'original' Report should be retained in a safe place and be shown to any person inspecting or undertaking work on the electrical installation in the future. If you later vacate the property, this Report will provide the new owner with details of the condition of the electrical installation at the time the Report was issued.

The 'Extent and Limitations' box should fully identify the extent of the installation covered by this Report and any limitations on the inspection and tests. The contractor should have agreed these aspects with you and with any other interested parties (Licensing Authority, Insurance Company, Building Society etc.) before the inspection was carried out.

The report should identify any departures from the safety requirements of the current Regulations and any defects, damage or deterioration that affect the safety of the installation for continued use. **For items classified as 'requires urgent attention', the safety of those using the installation may be at risk**, and it is recommended that a competent person undertakes the necessary remedial work without delay.

For safety reasons, the electrical installation will need to be re-inspected at appropriate intervals by a competent person. The maximum time interval recommended before the next inspection is stated in the Report under 'Next Inspection.'

Appendix 7

Form 3   Form No 124 /3

SCHEDULE OF INSPECTIONS

Methods of protection against electric shock

Both basic and fault protection:
- [N/A] (i) SELV (Note 1)
- [N/A] (ii) PELV
- [N/A] (iii) Double insulation (Note 2)
- [N/A] (iv) Reinforced insulation (Note 2)

Basic protection: (Note 3)
- [✓] (i) Insulation of live parts
- [X] (ii) Barriers or enclosures
- [N/A] (iii) Obstacles (Note 4)
- [N/A] (iv) Placing out of reach (Note 5)

Fault protection:
(i) Automatic disconnection of supply:
- [✓] Presence of earthing conductor
- [✓] Presence of circuit protective conductors
- [✓] Presence of protective bonding conductors
- [X] Presence of supplementary bonding conductors
- [✓] Presence of earthing arrangements for combined protective and functional purposes
- [N/A] Presence of adequate arrangements for alternative source(s), where applicable
- [N/A] FELV
- [✓] Choice and setting of protective and monitoring devices (for fault and/or overcurrent protection)

(ii) Non-conducting location: (Note 6)
- [N/A] Absence of protective conductors

(iii) Earth-free local equipotential bonding: (Note 6)
- [N/A] Presence of earth-free local equipotential bonding

(iv) Electrical Separation: (Note 7)
- [N/A] Provided for **one item** of current-using equipment
- [N/A] Provided for **more than one item** of current-using equipment

Additional protection:
- [X] Presence of residual current devices(s)
- [X] Presence of supplementary bonding conductors

Prevention of mutual detrimental influence
- [✓] (a) Proximity of non-electrical services and other influences
- [✓] (b) Segregation of Band I and Band II circuits or use of Band II insulation
- [✓] (c) Segregation of safety circuits

Identification
- [✓] (a) Presence of diagrams, instructions, circuit charts and similar information
- [✓] (b) Presence of danger notices and other warning notices
- [X] (c) Labelling of protective devices, switches and terminals
- [✓] (d) Identification of conductors

Cables and conductors
- [✓] Selection of conductors for current-carrying capacity and voltage drop
- [✓] Erection methods
- [✓] Routing of cables in prescribed zones
- [X] Cables incorporating earthed armour or sheath, or run within an earthed wiring system, or otherwise adequately protected against nails, screws and the like
- [X] Additional protection provided by 30 mA RCD for cables in concealed walls (where required in premises not under the supervision of a skilled or instructed person)
- [✓] Connection of conductors
- [✓] Presence of fire barriers, suitable seals and protection against thermal effects

General
- [✓] Presence and correct location of appropriate devices for isolation and switching
- [✓] Adequacy of access to switchgear and other equipment
- [X] Particular protective measures for special installations and locations
- [✓] Connection of single-pole devices for protection or switching in line conductors only
- [✓] Correct connection of accessories and equipment
- [N/A] Presence of undervoltage protective devices
- [✓] Selection of equipment and protective measures appropriate to external influences
- [✓] Selection of appropriate functional switching devices

Inspected by .... *S. Jones* ....   Date .... **26/08/08** ....

Notes:
- ✓ to indicate an inspection has been carried out and the result is satisfactory
- X to indicate an inspection has been carried out and the result is not satisfactory (applicable to a periodic inspection only)
- N/A to indicate the inspection is not applicable to a particular item
- LIM to indicate that, exceptionally, a limitation agreed with the person ordering the work prevented the inspection or test being carried out (applicable to a periodic inspection only).

1. SELV – an extra-low voltage system which is electrically separated from Earth and from other systems in such a way that a single-fault cannot give rise to the risk of electric shock. The particular requirements of the Regulations must be checked (see Section 414)
2. Double or reinforced insulation. Not suitable for domestic or similar installations if it is the sole protective measure (see 412.1.3)
3. Basic protection – will include measurement of distances where appropriate
4. Obstacles – only adopted in special circumstances (see 417.2)
5. Placing out of reach – only adopted in special circumstances (see 417.3)
6. Non-conducting locations and Earth-free local equipotential bonding – these are not recognised for general application. May only be used where the installation is controlled/under the supervision of skilled or instructed persons (see Section 418)
7. Electrical separation – the particular requirements of the Regulations must be checked. If a single item of current-using equipment is supplied from a single source, see Section 413. If more than one item of current-using equipment is supplied from a single source then the installation must be controlled/under the supervision of skilled or instructed persons, see also Regulation 418.3

Page 3 of 4

7 | Appendix

Form 4
SCHEDULE OF TEST RESULTS

Form No 126 /4

Contractor: County Electrics Ltd
Test Date: 26/08/08
Signature: W White
Method of fault protection: automatic disconnection of supply
Equipment vulnerable to testing: electric shower, discharge lamp

Address/Location of distribution board:
111 Any Street

Description of Work: Periodic inspection of house / garage / shed

*1 Type of Supply: ~~TN-S~~/TN-C-S/~~TT~~
*2 Ze at origin: 0.24 ohms
*3 PFC: 1.0 kA
Confirmation of supply polarity ☑

Instruments
loop impedance: LM 10
continuity: LM 11
insulation: LM 14
RCD tester: LM 16

Circuit Description	Overcurrent Device			Wiring Conductors		Continuity			Insulation Resistance		P o l a r i t y	Test Results			Remarks
		*4 Short-circuit capacity: 1.6.5 kA										Earth Loop Imped-ance	Functional Testing		
	type	Rating I_n		live	cpc	(R₁ + R₂)*	R₂*	R end of cct	Live/Live	Live/Earth		Z_s	RCD time	Other	
		A		mm²	mm²	Ω	Ω	Ω	MΩ	MΩ		Ω	ms		
1	2	3		4	5	*6	*7	*8	*9	*10	*11	*12	*13	*14	15
Lights up	1361	5		1.5	1.0	✓	–	–	10	8	✓	1.4	–	✓	pendants worn, lampholders O/H
Lights down	1361	5		1.5	1.0	✓	–	–	–	7	✓	1.3	–	✓	discharge lamp, lighting pendants worn
Ring up	"	30		2.5	1.5	✓	–	–	12	12	✓	0.5	–	✓	
Ring down	"	30		2.5	1.5	✓	–	–	13	13	✓	0.6	–	✓	broken socket
Cooker	"	30		6.0	2.5	✓	–	–	20	15	✓	0.3	–	✓	
Shower	"	30		6.0	2.5	✓	–	–	–	6	✓	0.4	–	✓	electric shower

Deviations from Wiring Regulations and special notes:
Main bonding cables 6mm² (not overheating)
No supplementary bonding to bathroom
Lighting pendants need replacing
No RCD to socket-circuits in house or garage or shed

No RCD to bathroom circuits
Broken socket in kitchen — accessible live parts
Consumer unit not labelled

*Complete column 6 or 7

*Number - See notes on schedule of test results on page 141 of Guide

Page 4 of 4

Appendix 8
Standard circuit arrangements for household and similar installations

8.1 Introduction

This appendix gives advice on standard circuit arrangements for household and similar premises. The circuits provide guidance on the requirements of Chapter 43 for overload protection and Section 537 of BS 7671 for isolation and switching. Reference must also be made to Section 7 and Table 7.1 for cable csa, length and installation reference method.

It is the responsibility of the designer and installer when adopting these circuit arrangements to take the appropriate measures to comply with the requirements of other chapters or sections which are relevant, such as Chapter 41 'Protection against electric shock', Chapter 54 'Earthing arrangements and protective conductors' and Chapter 52 'Selection and erection of wiring systems'.

Circuit arrangements other than those detailed in this appendix are not precluded when specified by a competent person, in accordance with the general requirements of Regulation 314.3.

8 Appendix

8.2 Final circuits using socket-outlets complying with BS 1363-2 and fused connection units complying with BS 1363-4

8.2.1 General

In this arrangement, a ring or radial circuit, with spurs if any, feeds permanently connected equipment and a number of socket-outlets and fused connection units.

The floor area served by the circuit is determined by the known or estimated load and should not exceed the value given in Table 8A.

433.1.5 A single 30 A or 32 A ring circuit may serve a floor area of up to 100 m². Sockets for washing machines, tumble dryers and dishwashers should be located so as to provide reasonable sharing of the load in each leg of the ring, or consideration should be given to separate circuits.

553.1.7 The number of socket-outlets provided should be such that all equipment can be supplied from an adjacent accessible socket-outlet, taking account of the length of flex normally fitted to portable appliances and luminaires.

Diversity between socket-outlets and permanently connected equipment has already been taken into account in Table 8A and no further diversity should be applied, see Appendix 1 of this Guide.

▼ **Table 8A** Final circuits using BS 1363 socket-outlets and connection units

			Minimum live conductor cross-sectional area* (mm²)		
Type of circuit		Overcurrent protective device rating (A)	Copper conductor thermoplastic or thermosetting insulated cables	Copper conductor mineral insulated cables	Maximum floor area served (m²)
1	2	3	4	5	6
A1	Ring	30 or 32	2.5	1.5	100
A2	Radial	30 or 32	4	2.5	75
A3	Radial	20	2.5	1.5	50

* See Section 7 and Table 7.1 for the minimum csa for particular installation reference methods. It is permitted to reduce the values of conductor cross-sectional area for fused spurs.

Where two or more ring final circuits are installed, the socket-outlets and permanently connected equipment to be served should be reasonably distributed among the circuits.

Appendix | 8

8.2.2 Circuit protection

Table 8A is applicable for circuits protected by:

- fuses to BS 3036, BS 1361 and BS 88, and
- circuit-breakers:
  - Types B and C to BS EN 60898 or BS EN 61009-1
  - BS EN 60947-2
  - Types 1, 2 and 3 to BS 3871.

8.2.3 Conductor size

The minimum size of conductor cross-sectional area in the circuit and in non-fused spurs is given in Table 8A. However, the actual size of cable is determined by the current-carrying capacity for the particular method of installation, after applying appropriate rating factors from Appendix 6, see Table 7.1. The as-installed current-carrying capacity (I_z) so calculated must be not less than:

- 20 A for ring circuit A1,
- 30 A or 32 A for radial circuit A2 (i.e. the rating of the overcurrent protective device),
- 20 A for radial circuit A3 (i.e. the rating of the overcurrent protective device).

The conductor size for a fused spur is determined from the total current demand served by that spur, which is limited to a maximum of 13 A.

Where a fused spur serves socket-outlets the minimum conductor size is:

- 1.5 mm^2 for cables with thermosetting or thermoplastic insulated cables, copper conductors,
- 1 mm^2 for mineral insulated cables, copper conductors.

The conductor size for circuits protected by BS 3036 fuses is determined by applying the 0.725 factor of Regulation 433.1.3, that is the current-carrying capacity must be at least 27 A for circuits A1 and A3, 41 A for circuit A2.

8.2.4 Spurs

The total number of fused spurs is unlimited but the number of non-fused spurs should not exceed the total number of socket-outlets and items of stationary equipment connected directly in the circuit.

In an A1 ring final circuit and an A2 radial circuit of Table 8A a non-fused spur should feed only one single or one twin or multiple socket-outlet or one item of permanently connected equipment. Such a spur should be connected to the circuit at the terminals of a socket-outlet or junction box or at the origin of the circuit in the distribution board.

A fused spur should be connected to the circuit through a fused connection unit, the rating of the fuse in the unit not exceeding that of the cable forming the spur and, in any event, not exceeding 13 A.

8 Appendix

8.2.5 Permanently connected equipment

537.5.1.3
Table 53.2

Permanently connected equipment should be locally protected by a fuse complying with BS 1362 of rating not exceeding 13 A or by a circuit-breaker of rating not exceeding 16 A and should be controlled by a switch, where needed (see Appendix 10). A separate switch is not required if the circuit-breaker is to be used as a switch.

8.3 Radial final circuits using 16 A socket-outlets complying with BS EN 60309-2 (BS 4343)

8.3.1 General

Where a radial circuit feeds equipment the maximum demand of which, having allowed for diversity, is known or estimated not to exceed the rating of the overcurrent protective device and in any event does not exceed 20 A, the number of socket-outlets is unlimited.

8.3.2 Circuit protection

The overcurrent protective device should have a rating not exceeding 20 A.

8.3.3 Conductor size

The minimum size of conductor in the circuit is given in Tables 8A and 7.1. Where cables are grouped together the limitations of 7.2.1 and Appendix 6 apply.

8.3.4 Types of socket-outlet

Socket-outlets should have a rated current of 16 A and be of the type appropriate to the number of phases, circuit voltage and earthing arrangements. Socket-outlets incorporating pilot contacts are not included.

8.4 Cooker circuits in household and similar premises

The circuit supplies a control switch or a cooker unit complying with BS 4177, which may incorporate a socket-outlet.

The rating of the circuit is determined by the assessment of the current demand of the cooking appliance(s), and cooker control unit socket-outlet if any, in accordance with Table 1A of Appendix 1. A 30 or 32 A circuit is usually appropriate for household or similar cookers of rating up to 15 kW.

A circuit of rating exceeding 15 A but not exceeding 50 A may supply two or more cooking appliances where these are installed in one room. The control switch or cooker control unit should be placed within 2 m of the appliance, but not directly above it. Where two stationary cooking appliances are installed in one room, one switch may be used to control both appliances provided that neither appliance is more than 2 m from the switch. Attention is drawn to the need to provide selective (discriminative) operation of protective devices as stated in Regulation 536.2.

Appendix | 8

8.5 Water and space heating

Water heaters fitted to storage vessels in excess of 15 litres capacity, or permanently connected heating appliances forming part of a comprehensive space heating installation, should be supplied by their own separate circuit.

Immersion heaters should be supplied through a switched cord-outlet connection unit complying with BS 1363-4.

8.6 Height of switches, socket-outlets and controls

The Building Regulations require switches and socket-outlets in new dwellings to be installed so that all persons including those whose reach is limited can easily use them. A way of satisfying the requirement is to install switches, socket-outlets and controls throughout the dwelling in accessible positions at a height of between 450 mm and 1200 mm from the finished floor level – see Figure 8A. Because of the sensitivity of circuit-breakers, RCCBs and RCBOs fitted in consumer units, consumer units should be readily accessible.

553.1.6

(In areas subject to flooding, meters, cut-outs and consumer units should preferably be fixed above flood water level.)

▼ **Figure 8A**   Height of switches, sockets, etc. (see Approved Document M, section 8)

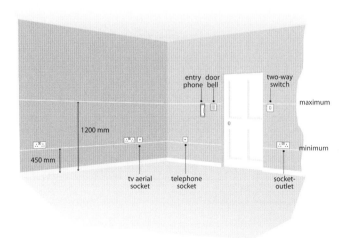

8 Appendix

8.7 Number of socket-outlets

553.1.7 Sufficient socket-outlets are required to be installed so that all equipment likely to be used can be supplied from a reasonably accessible socket-outlet, taking account of the length of flexible cords normally fitted to portable appliances and luminaires. Table 8B provides guidance on the number of socket-outlets that are likely to meet this requirement.

In Scotland, mandatory standard 4.6 requires that every building must be designed and constructed in such a way that electric lighting points and socket-outlets are provided to ensure the health, safety and convenience of occupants and visitors. The Scottish Building Standards Agency (SBSA) make recommendations for the number of socket-outlets that should be installed in a domestic premises in section 4.6.4 of the domestic technical handbook as follows:

- kitchen – 6 (at least 3 above worktop height)
- other habitable rooms – 4
- plus at least 4 more throughout the property including at least one per circulation area per storey.

The socket-outlets may be either single or double.

▼ **Table 8B** Minimum number of twin socket-outlets to be provided in homes

Room type	Smaller rooms (up to 12 m²)	Medium rooms (12–25 m²)	Larger rooms (more than 25 m²)
Main living room (note 4)	4	6	8
Dining room	3	4	5
Single bedroom (note 3)	2	3	4
Double bedroom (note 3)	3	4	5
Bedsitting room (note 6)	4	5	6
Study	4	5	6
Utility room	3	4	5
Kitchen (note 1)	6	8	10
Garage (note 2)	2	3	4
Conservatory	3	4	5
Hallway	1	2	3
Loft	1	2	3
Location containing a bath or shower		note 5	

Appendix 8

Notes to Table 8B:

1. **KITCHEN** – If a socket-outlet is provided in the cooker control unit, this should not be included in the 6 recommended in the table above.
Appliances built into kitchen furniture (integrated appliances) should be connected to a socket-outlet or switch fused connection unit that is accessible when the appliance is in place and in normal use. Alternatively, where an appliance is supplied from a socket-outlet or a connection unit, these should be controlled by an accessible double-pole switch or switched fused connection unit.
It is recommended that wall mounted socket-outlets above a work surface are spaced at not more than 1 m intervals along the surface.

2. **GARAGE** – The number of socket-outlets specified allows for the use of a battery charger, tools, portable light and garden appliances.

3. **BEDROOM** – It is envisaged that this room will be used in different ways in different households. It may be used simply as a child's bedroom requiring socket-outlets for table lamps, an electric blanket and an electric heater only; or it may serve as a teenager's bedroom and living room combined, where friends are entertained. In this case, socket-outlets may be needed for computers (printers, scanners, speakers, etc.), games consoles, MP3/4 players, mobile phone chargers, DVD players, digital receivers, home entertainment systems (amplifier, CD player), hairdryer, television and radio, in addition to lamps, an electric blanket and electric heater.

4. **HOME ENTERTAINMENT** – In addition to the number of socket-outlets shown in the table it is recommended that at least two further double socket-outlets are installed in home entertainment areas.

5. **LOCATIONS CONTAINING A BATH OR SHOWER** – Except for SELV socket-outlets complying with Section 414 and shaver supply units complying with BS EN 61558-2-5, socket-outlets are prohibited within a distance of 3 m horizontally from the boundary of zone 1.

6. **BEDSITTING ROOM** – Rooms specifically designed or envisaged to be used as student bedsitting rooms should be provided with additional socket-outlets which may be needed since persons using these rooms will often introduce other portable appliances in addition to items already mentioned in Note 3. In such situations a lack of sufficient socket-outlets is an additional danger and therefore the minimum number of twin outlets should be increased to four.

Appendix 9
Resistance of copper and aluminium conductors

To check compliance with Regulation 434.5.2 and/or Regulation 543.1.3, i.e. to evaluate the equation $S^2 = I^2.t/k^2$, it is necessary to establish the impedances of the circuit conductors to determine the fault current I and hence the protective device disconnection time t.

434.5.2
543.1.3

Fault current $I = U_0/Z_s$

where:

U_0 is the nominal voltage to earth,
Z_s is the earth fault loop impedance.

$Z_s = Z_e + R_1 + R_2$

where:

Z_e is that part of the earth fault loop impedance external to the circuit concerned,
R_1 is the resistance of the line conductor from the origin of the circuit to the point of utilization,
R_2 is the resistance of the protective conductor from the origin of the circuit to the point of utilization.

Similarly, in order to design circuits for compliance with BS 7671 limiting values of earth fault loop impedance given in Tables 41.2, 41.3 and 41.4, it is necessary to establish the relevant impedances of the circuit conductors concerned at their operating temperature.

Table 9A gives values of $(R_1 + R_2)$ per metre for various combinations of conductors up to and including 35 mm^2 cross-sectional area. It also gives values of resistance (milliohms) per metre for each size of conductor. These values are at 20 °C.

9 Appendix

▼ **Table 9A** Values of resistance/metre or $(R_1 + R_2)$/metre for copper and aluminium conductors at 20 °C

Cross-sectional area (mm²)		Resistance/metre or $(R_1 + R_2)$/metre (mΩ/m)	
Line conductor	Protective conductor	Copper	Aluminium
1	–	18.10	
1	1	36.20	
1.5	–	12.10	
1.5	1	30.20	
1.5	1.5	24.20	
2.5	–	7.41	
2.5	1	25.51	
2.5	1.5	19.51	
2.5	2.5	14.82	
4	–	4.61	
4	1.5	16.71	
4	2.5	12.02	
4	4	9.22	
6	–	3.08	
6	2.5	10.49	
6	4	7.69	
6	6	6.16	
10	–	1.83	
10	4	6.44	
10	6	4.91	
10	10	3.66	
16	–	1.15	1.91
16	6	4.23	–
16	10	2.98	–
16	16	2.30	3.82
25	–	0.727	1.20
25	10	2.557	–
25	16	1.877	–
25	25	1.454	2.40
35	–	0.524	0.87
35	16	1.674	2.78
35	25	1.251	2.07
35	35	1.048	1.74
50	–	0.387	0.64
50	25	1.114	1.84
50	35	0.911	1.51
50	50	0.774	1.28

Appendix 9

▼ **Table 9B**   Ambient temperature multipliers to Table 9A

Expected ambient temperature (°C)	Correction factor*
5	0.94
10	0.96
15	0.98
20	1.00
25	1.02

* The correction factor is given by {1 + 0.004(ambient temp − 20 °C)}
where 0.004 is the simplified resistance coefficient per °C at 20 °C given by BS EN 60228 for copper and aluminium conductors.

Verification

For verification purposes the designer will need to give the values of the line and circuit protective conductor resistances at the ambient temperature expected during the tests. This may be different from the reference temperature of 20 °C used for Table 9A. The rating factors in Table 9B may be applied to the values to take account of the ambient temperature (for test purposes only).

Multipliers for conductor operating temperature

Table 9C gives the multipliers to be applied to the values given in Table 9A for the purpose of calculating the resistance at maximum operating temperature of the line conductors and/or circuit protective conductors in order to determine compliance with, as applicable, the earth fault loop impedance of Table 41.2, 41.3 or 41.4 of BS 7671.

Table 41.2
Table 41.3
Table 41.4

Where it is known that the actual operating temperature under normal load is less than the maximum permissible value for the type of cable insulation concerned (as given in the tables of current-carrying capacity) the multipliers given in Table 9C may be reduced accordingly.

9 Appendix

▼ **Table 9C** Multipliers to be applied to Table 9A to calculate conductor resistance at maximum operating temperature (note 3) for standard devices (note 4)

Conductor installation	Conductor insulation		
	70 °C Thermoplastic (PVC)	90 °C Thermoplastic (PVC)	90 °C Thermosetting
Not incorporated in a cable and not bunched (note 1)	1.04	1.04	1.04
Incorporated in a cable or bunched (note 2)	1.20	1.26	1.28

Notes:

Table 54.2   1   See Table 54.2 of BS 7671, which applies where the protective conductor is not incorporated or bunched with cables, or for bare protective conductors in contact with cable covering.
Table 54.3   2   See Table 54.3 of BS 7671, which applies where the protective conductor is a core in a cable or is bunched with cables.
3   The multipliers given in Table 9C for both copper and aluminium conductors are based on a simplification of the formula given in BS EN 60228, namely that the resistance–temperature coefficient is 0.004 per °C at 20 °C.
4   Standard devices are those described in Appendix 3 of BS 7671 (fuses to BS 1361, BS 88, BS 3036, circuit-breakers to BS EN 60898 types B, C, and D) and BS 3871-1.

Appendix 10

Selection of devices for isolation and switching

▼ **Table 10A** Summary of the functions provided by devices for isolation and switching   Table 53.2

Device	Standard	Isolation[5]	Emergency switching[2,5]	Functional switching[5]
Switching device	BS 3676: Pt 1 1989	Yes[4]	Yes	Yes
	BS EN 60669-1	No	Yes	Yes
	BS EN 60669-2-1	No	No	Yes
	BS EN 60669-2-2	No	Yes	Yes
	BS EN 60669-2-3	No	Yes	Yes
	BS EN 60669-2-4	Yes	Yes	Yes
	BS EN 60947-3	Yes[1]	Yes	Yes
	BS EN 60947-5-1	No	Yes	Yes
Contactor	BS EN 60947-4-1	Yes[1]	Yes	Yes
	BS EN 61095	No	No	Yes
Circuit-breaker	BS EN 60898	Yes	Yes	Yes
	BS EN 60947-2	Yes[1]	Yes	Yes
	BS EN 61009-1	Yes	Yes	Yes
RCD	BS EN 60947-2	Yes[1]	Yes	Yes
	BS EN 61008-1	Yes	Yes	Yes
	BS EN 61009-1	Yes	Yes	Yes
Isolating switch	BS EN 60669-2-4	Yes	Yes	Yes
	BS EN 60947-3	Yes	Yes	Yes
Plug and socket-outlet ($\leq$ 32 A)	BS EN 60309	Yes	No	Yes
	IEC 60884	Yes	No	Yes
	IEC 60906	Yes	No	Yes
Plug and socket-outlet ($>$ 32 A)	BS EN 60309	Yes	No	No
Device for the connection of luminaire	BS IEC 61995-1	Yes[3]	No	No

continues

10 Appendix

▼ **Table 10A** *continued*

Device	Standard	Isolation[5]	Emergency switching[2,5]	Functional switching[5]
Control and protective switching device for equipment (CPS)	BS EN 60947-6-1 BS EN 60947-6-2	Yes Yes[1]	Yes Yes	Yes Yes
Fuse	BS 88	Yes	No	No
Device with semiconductors	BS EN 60669-2-1	No	No	Yes
Luminaire-supporting coupler	BS 6972	Yes[3]	No	No
Plug and unswitched socket-outlet	BS 1363-1 BS 1363-2	Yes[3] Yes[3]	No No	Yes Yes
Plug and switched socket-outlet	BS 1363-1 BS 1363-2	Yes[3] Yes[3]	No No	Yes Yes
Plug and socket-outlet	BS 5733	Yes[3]	No	Yes
Switched fused connection unit	BS 1363-4	Yes[3]	Yes	Yes
Unswitched fused connection unit	BS 1363-4	Yes[3] (removal of fuse link)	No	No
Fuse	BS 1362	Yes	No	No
Cooker control unit switch	BS 4177	Yes	Yes	Yes

Notes:
1. Function provided if the device is suitable and marked with the symbol for isolation (see BS EN 60617 identity number S00288). ─/─
2. The means of operation shall be readily accessible at places where a danger might occur and, where appropriate, at any additional remote position from which that danger can be removed. See Regulation 537.4.2.5.
3. Device is suitable for on-load isolation, i.e. disconnection whilst carrying load current.
4. Function provided if the device is suitable and marked with ◯.
5. 'Yes' indicates function provided, 'No' indicates function not provided.

Appendix 11
Identification of conductors

11.1 Introduction
Appx 7

The requirements of BS 7671 have been harmonized with the technical intent of CENELEC Standard HD 384.5.514: *Identification*, including 514.3: *Identification of conductors*.

Amendment No. 2:2004 (AMD 14905) to BS 7671 implemented the harmonized cable core colours and the alphanumeric marking of the following standards:

- ▶ HD 308 S2:2001 *Identification of cores in cables and flexible cords*
- ▶ BS EN 60445:2000 *Basic and safety principles for man–machine interface, marking and identification of equipment and terminals and of terminations*
- ▶ BS EN 60446:2000 *Basic and safety principles for man–machine interface, marking and identification of equipment by colours or numerals*

This appendix provides guidance on marking at the interface between old and harmonized colours, and general guidance on the colours to be used for conductors.

British Standards for fixed and flexible cables have been harmonized (see Table 11A). BS 7671 has been modified to align with these cables, but also allows other suitable methods of marking connections by colours (tapes, sleeves or discs), or by alphanumerics (letters and/or numbers). Methods may be mixed within an installation.

11 Appendix

Table 51
514.3.1

▼ **Table 11A** Identification of conductors (harmonized)

Function	Alphanumeric	Colour
Protective conductors		Green-and-Yellow
Functional earthing conductor		Cream
a.c. power circuit[1]		
Line of single-phase circuit	L	Brown
Neutral of single- or three-phase circuit	N	Blue
Line 1 of three-phase a.c. circuit	L1	Brown
Line 2 of three-phase a.c. circuit	L2	Black
Line 3 of three-phase a.c. circuit	L3	Grey
Two-wire unearthed d.c. power circuit		
Positive of two-wire circuit	L+	Brown
Negative of two-wire circuit	L-	Grey
Two-wire earthed d.c. power circuit		
Positive (of negative earthed) circuit	L+	Brown
Negative (of negative earthed) circuit[2]	M	Blue
Positive (of positive earthed) circuit[2]	M	Blue
Negative (of positive earthed) circuit	L-	Grey
Three-wire d.c. power circuit		
Outer positive of two-wire circuit derived from three-wire system	L+	Brown
Outer negative of two-wire circuit derived from three-wire system	L-	Grey
Positive of three-wire circuit	L+	Brown
Mid-wire of three-wire circuit[2,3]	M	Blue
Negative of three-wire circuit	L-	Grey
Control circuits, ELV and other applications		
Line conductor	L	Brown, Black, Red, Orange, Yellow, Violet, Grey, White, Pink or Turquoise
Neutral or mid-wire[4]	N or M	Blue

Notes:
1. Power circuits include lighting circuits.
2. M identifies either the mid-wire of a three-wire d.c. circuit, or the earthed conductor of a two-wire earthed d.c. circuit.
3. Only the middle wire of three-wire circuits may be earthed.
4. An earthed PELV conductor is blue.

Appendix 11

11.2 Addition or alteration to an existing installation

11.2.1 Single-phase

An addition or alteration made to a single-phase installation need not be marked at the interface provided that:

i the old cables are correctly identified by the colour red for line and black for neutral, and
ii the new cables are correctly identified by the colour brown for line and blue for neutral.

11.2.2 Two- or three-phase installation

Where an addition or alteration is made to a two- or a three-phase installation wired in the old core colours with cable to the new core colours, unambiguous identification is required at the interface. Cores shall be marked as follows:

Neutral conductors
Old and new conductors: N

Line conductors
Old and new conductors: L1, L2, L3

▼ **Table 11B** Example of conductor marking at the interface for additions and alterations to an a.c. installation identified with the old cable colours Table 7A

Function	Old conductor		New conductor	
	Colour	Marking	Marking	Colour
Line 1 of a.c.	Red	L1	L1	Brown*
Line 2 of a.c.	Yellow	L2	L2	Black*
Line 3 of a.c.	Blue	L3	L3	Grey*
Neutral of a.c.	Black	N	N	Blue
Protective conductor	Green-and-Yellow			Green-and-Yellow

* Three single-core cables with insulation of the same colour may be used if identified at the terminations.

11.3 Switch wires in a new installation or an addition or alteration to an existing installation

Where a two-core cable with cores coloured brown and blue is used as a switch wire, both conductors being line conductors, the blue conductor shall be marked brown or L at its terminations.

On-Site Guide **173**
© The Institution of Engineering and Technology

11 Appendix

11.4 Intermediate and two-way switch wires in a new installation or an addition or alteration to an existing installation

Where a three-core cable with cores coloured brown, black and grey is used as a switch wire, all three conductors being line conductors, the black and grey conductors shall be marked brown or L at their terminations.

11.5 Line conductors in a new installation or an addition or alteration to an existing installation

Power circuit line conductors should be coloured as in Table 11A. Other line conductors may be brown, black, red, orange, yellow, violet, grey, white, pink or turquoise.

In a two- or three-phase power circuit the line conductors may all be of one of the permitted colours, either identified L1, L2, L3 or marked brown, black, grey at their terminations to show the phases.

11.6 Changes to cable core colour identification

Table 7B ▼ **Table 11C**   Cable to BS 6004 (flat cable with bare cpc)

Cable type	Old core colours	New core colours
Single-core + bare cpc	Red or Black	Brown or Blue
Two-core + bare cpc	Red, Black	Brown, Blue
Alt. two-core + bare cpc	Red, Red	Brown, Brown
Three-core + bare cpc	Red, Yellow, Blue	Brown, Black, Grey

Table 7C ▼ **Table 11D**   Standard 600/1000 V armoured cable BS 6346, BS 5467 or BS 6724

Cable type	Old core colours	New core colours
Single-core	Red or Black	Brown or Blue
Two-core	Red, Black	Brown, Blue
Three-core	Red, Yellow, Blue	Brown, Black, Grey
Four-core	Red, Yellow, Blue, Black	Brown, Black, Grey, Blue
Five-core	Red, Yellow, Blue, Black, Green-and-Yellow	Brown, Black, Grey, Blue, Green-and-Yellow

Appendix 11

▼ **Table 11E**   Flexible cable to BS 6500   Table 7D

Cable type	Old core colours	New core colours
Two-core	Brown, Blue	No change
Three-core	Brown, Blue, Green-and-Yellow	No change
Four-core	Black, Blue, Brown, Green-and-Yellow	Brown, Black, Grey, Green-and-Yellow
Five-core	Black, Blue, Brown, Black, Green-and-Yellow	Brown, Black, Grey, Blue, Green-and-Yellow

11.7   Addition or alteration to a d.c. installation

Where an addition or alteration is made to a d.c. installation wired in the old core colours with cable to the new core colours, unambiguous identification is required at the interface. Cores shall be marked as follows:

Neutral and midpoint conductors
Old and new conductors: M

Line conductors
Old and new conductors: Brown or Grey, or L+ or L-

▼ **Table 11F**   Example of conductor marking at the interface for additions and   Table 7E
alterations to a d.c. installation identified with the old cable colours

Function	Old conductor		New conductor	
	Colour	Marking	Marking	Colour
Two-wire unearthed d.c. power circuit				
Positive of two-wire circuit	Red	L+	L+	Brown
Negative of two-wire circuit	Black	L-	L-	Grey
Two-wire earthed d.c. power circuit				
Positive (of negative earthed) circuit	Red	L+	L+	Brown
Negative (of negative earthed) circuit	Black	M	M	Blue
Positive (of positive earthed) circuit	Black	M	M	Blue
Negative (of positive earthed) circuit	Blue	L-	L-	Grey
Three-wire d.c. power circuit				
Outer positive of two-wire circuit derived from three-wire system	Red	L+	L+	Brown
Outer negative of two-wire circuit derived from three-wire system	Red	L-	L-	Grey
Positive of three-wire circuit	Red	L+	L+	Brown
Mid-wire of three-wire circuit	Black	M	M	Blue
Negative of three-wire circuit	Blue	L-	L-	Grey

Index

A
Additional protection	
RCDs	3.6.1
supplementary bonding	4.6
Alarms, smoke and heat	7.5.2
Alphanumeric identification	Table 11A
Automatic disconnection	3.5

B
Band I	7.4.1
Band II	7.4.1
Basic protection	3.4.1
Bathrooms	4.6; 7.2.5; 8.1
Bends, cable	Table 4E
Bonding	4
BS 1363 socket-outlets	Appx 8
Building Regulations	1.2; Fig 8A

C
Cable	
bends	Table 4E
floors and ceilings	7.3.1
installation methods	Table 7.1; Table 7.2; Table 7.3
ratings	Appx 6
selection	Appx 3
spans (overhead wiring)	Table 4B
supports	Appx 4
walls and partitions	7.3.2
Capacities	
conduits	Appx 5
trunking	Appx 5
Ceilings	7.3.1
Central heating	4.2
Certificates	9.1; Appx 7
Checklist	
inspection	9.2.2
testing	9.3.1
Circuit arrangement	3.6.2; Appx 8
Circuit-breaker selection	Table 7.5
Circuit protective conductors	Fig 2.1; Fig 2.2; 3.4.2
Circuits	7
Colours, cable core	Appx 11
Competent person	Preface; Foreword
Conduit	
capacities	Appx 5
supports	Table 4C
Consumer unit	2.2.5; 3.3; Fig 3.1; Fig 3.2; Fig 3.3; Fig 3.4; Fig 3.5
Consumer's controlgear	2.2.5
Continuity testing	9.3; 10.3.1
of rings	10.3.2
Cooker circuit	Table 1A; Table 1B; Appx 8.4
Corrosion	Appx 3
Current-carrying capacity	Appx 6
Cut-out	1.1(iii); 2.2.1

D
d.c.	Appx 11.7
Devices, selection of	Appx 10
Diagrams	6.1(x)
Direct contact	3.4.1
Direct current	Appx 11.7
Disconnection times	3.5; 7.2.7(iv)
Distribution board	3.1
Distributor's cut-out	1.1(iii); 2.2.1
Diversity	Appx 1
Dual supply, notice	Fig 6.5

E
Earth electrode	4.8
testing	10.3.5
types	4.9

176 | On-Site Guide
© The Institution of Engineering and Technology

Index

Earthing	4
conductor size	4.3
TN-C-S	Fig 2.1
TN-S	Fig 2.2
TT	Fig 2.3
Earth fault loop impedance	
circuits	Appx 2
RCD	3.6.1(i) supply
1.1(iv)	
testing	9.3; 10.3.6
Electrical installation certificates	Appx 7
Electric shock	3.4; 8.1
Emergency switching	5.3
Equipotential bonding	4

F

Fault current	
measurement	10.3.7
protection	3.3; 7.2.7(i)
Fault protection	3.4.2
Final circuits	7
Fire alarms	7.5.2
Firefighter's switch	5.5
Fixed wiring	Table 3A
Flexible cords	Table 3B
Floors	7.3.1
Functional	
extra-low voltage	10.3.3(vi)
switching	5.4
testing	10.3.9
Fuseboard	3.3
Fuses	7.2.7(ii)
distributor's	1.1(iii); 2.2.1
types, selection	7.2.7

G

Gas pipes	2.3; 4.3
Gas service	4.2
Grouping	7.2.1; Table 6C

H

Hearing aid loop	7.4.4
Heat alarms	7.5.2
Height of overhead wiring	Table 4B
Height of switches, sockets	Fig 8A
High protective conductor current	7.6

I

Identification of conductors	Appx 11
Immersion heaters	Appx 8.5

Indirect contact	3.4.1
Induction loop	7.4.4
Information	1.3
Inspection and testing	9
Inspection report	Appx 7
Inspection schedule	9.1; Appx 7
Installation method	7.1(iii)
Insulation resistance	9.3.1; 10.3.3
Isolation	5.1

J

Joists	7.3.1

L

Labelling	6
Length of span, overhead wiring	Table 4B
Lighting circuits	7.2.3; Table 7.1
Lighting demand	Table 1A
diversity	Table 1B
Lightning protection, bonding	4.2
Line conductor	1.1
Live part	1.1
Load characteristics	7.2.7(ii)
Load estimation	Appx 1
Loop impedance	Appx 2
testing	10.3.6

M

Main earthing	4
bonding	Fig 2.1; Fig 2.2; Fig 2.3
terminal	Fig 2.1; Fig 2.2; Fig 2.3
Main switch	Fig 2.1; Fig 2.2; Fig 2.3
Maximum demand	Appx 1
Maximum length of circuit	Table 7.1
Mechanical maintenance	5.2
Metal pipework	4
Metal structures	4.2
Meter	2.2.2
Meter tails	2.2.3; 4.10
Mineral cable	Table 3A
Minor works certificate	Appx 7
Mobile equipment	3.6.1(iv)
Motors	
circuit-breakers	Table 7.5
diversity	Table 1B
fuses	7.2.7(ii)(c)

N

Notices	6
Number of socket-outlets	Table 8B

Index

O
Oil supply pipe	4.2
Overhead wiring	Appx 4
Overload protection	3.2

P
Partition walls	7.3.2
Part P	1.2
PELV	3.4.3; 9.2.2; Table 10.1; 10.3.3(v)
Periodic reports	Appx 7
Phase sequence check	10.3.8
Plastic pipes	4.4; 4.7
Polarity testing	9.3; 10.3.4
Portable	3.6.1(iv)
Protection	3
Protective bonding	4
Protective conductors	4
Protective conductor current	7.6
Protective device	3.1; 7.2.7
Protective earthing	4.1
Proximity	7.4

R
Radial circuits	Appx 8
testing	10.3.1
RCBOs	3.6.2(c)
RCDs	3.6; 7.2.4
RCD testing	11
Reference (installation) method	7.1(iii)
Reports	Appx 7
Resistance of conductors	Appx 9
Ring circuits	Appx 8
testing	10.3.2
Rod, earth	Fig 2.3

S
Schedules	9.1; Appx 7
Scope	1.1
Segregation of circuits	7.4.1
Selection	
cables and cords	Appx 3
devices for isolation, etc.	Appx 10
SELV	3.4.3; 9.2.2; Table 10.1; 10.3.3(v)
Sequence of tests	10.2
Service position	2
Short-circuit protection	3.3; Table 7.4
Showers	4.6; 7.2.5; 8
Skilled persons	1.1
Smoke alarms	7.5.2
Socket-outlets	3.6.1; 7.2.2; Appx 8
height of	Fig 8A
Specification	Foreword
Spurs	Appx 8
Standard circuits	7.2; Appx 8
Statutory regulations	Preface
Supplementary bonding	4.5; 4.6; 4.7
Supplier's cut-out	1.1(iii); 2.2.1
Supplier's switch	2.2.4
Supply	1.1
Supply tails	4.10
Support, methods of	Appx 4
Switching	5
Switches, height of	Fig 8A

T
Telecommunication circuits	7.4.2
Test equipment	10.1
Test results schedule	9.1; Appx 7
Testing	9; 10; 11
Thermal insulation	Appx 6
Thermoplastic (PVC) cable	Table 7.1; Table 7.2; Table 6D1; Table 6E1; Table 6F
Thermosetting cable	Table 6D1; Table 6E1
TN-C-S system	Fig 2.1
TN-S system	Fig 2.2
Trunking	
capacities	Appx 5
supports	Table 4D
TT system	Fig 2.3; 7.2.5
Two-way circuits	Fig 7.3; 10.3.3

U
Underfloor heating	8.2

V
Voltage bands	7.4.1
Voltage drop	7.1; 7.2.3; Appx 6; Table 6D2; Table 6E2; Table 6F

W
Walls	7.3.2
Warning notices	6
Water heaters	Appx 8.5
Water pipes	4.2; 4.8

Errata

This list of changes is provided to help trainers and lecturers and those who wish to update earlier printed versions of the book.

October 2008 reprint

Page 59 (Section 7.3.1)
Add 'or' at end of alternatives i and ii.

Pages 60 (Section 7.3.2) and 61 (Section 7.4.1)
Add 'or' at end of alternatives i, ii and iii.

Page 92 (Section 11.6)
Delete ',' from end of alternative **a** and ', or' from end of alternative **b**.

Page 99 (Line 7)
Replace 'lsf' with 'low smoke halogen-free – LSHF'.

Pages 100 (Line 9) and 107 (Line 29)
Replace 'lsf' with 'LSHF'.

Page 103/4 (Table 2D)
The second panel has been amended to include more valid and usable information. Delete the entire second panel and its note, under the sub-heading **Minimum protective conductor size (mm²)**, and replace with the following.

Regulation 434.5.2 of BS 7671:2008 requires that the protective conductor csa meets the requirements of BS EN 60898-1, -2 or BS EN 61009-1, or the minimum quoted by the manufacturer. The values below are for energy limiting class 3, type B and C devices only.

Energy limiting class 3 device rating	Fault level (kA)	Protective conductor csa (mm²)	
		Type B	Type C
Up to and including 16 A	≤ 3	1.0	1.5
Up to and including 16 A	≤ 6	2.5	2.5
Over 16 up to and including 32 A	≤ 3	1.5	1.5
Over 16 up to and including 32 A	≤ 6	2.5	2.5
40 A	≤ 3	1.5	1.5
40 A	≤ 6	2.5	2.5

* For other device types and ratings or higher fault levels, consult manufacturer's data. See Regulation 434.5.2 and the IET publication *Commentary on the IEE Wiring Regulations*.

Errata

Page 106 (Note 2)
Replace 'BS 7540:1994' with 'BS 7540:2005 (series) *Electric cables — Guide to use for cables with a rated voltage not exceeding 450/750 V'*.

Pages 107 (Note 5) and 109 (Note 3)
Replace 'BS 7540:1994' with 'BS 7540:2005 (series)'.

IEE Wiring Regulations and associated publications

The IEE prepares regulations for the safety of electrical installations for buildings, the *IEE Wiring Regulations* (BS 7671: *Requirements for Electrical Installations*), which have now become the standard for the UK and many other countries. It also recommends, internationally, the requirements for ships and offshore installations. The IEE provides guidance on the application of the installation regulations through publications focused on the various activities from design of the installation through to final test and then maintenance. This includes a series of eight Guidance Notes, two Codes of Practice and Model Forms for use in Wiring Installations.

Requirements for Electrical Installations BS 7671:2008 (IEE Wiring Regulations, 17th Edition)
Order book PWR1700B Paperback 2008
ISBN: 978-0-86341-844-0 **£65**

On-Site Guide (BS 7671:2008 17th Edition)
Order book PWGO170B Paperback 2008
ISBN: 978-0-86341-854-9 **£20**

Wiring Matters Magazine FREE
If you wish to receive a FREE copy or advertise in Wiring Matters please visit www.theiet.org/wm

IEE Guidance Notes

A series of Guidance Notes has been issued, each of which enlarges upon and amplifies the particular requirements of a part of the IEE Wiring Regulations.

Guidance Note 1: Selection & Erection of Equipment, 5th Edition
Order book PWG1170B 237pp Paperback 2008
ISBN: 978-0-86341-855-6 **£30**

Guidance Note 2: Isolation & Switching, 5th Edition
Order book PWG2170B 74pp Paperback 2008
ISBN: 978-0-86341-856-3 **£25**

Guidance Note 3: Inspection & Testing, 5th Edition
Order book PWG3170B 126pp Paperback 2008
ISBN: 978-0-86341-857-0 **£25**

Guidance Note 4: Protection Against Fire, 5th Edition
Order book PWG4170B 98pp Paperback 2008
ISBN: 978-0-86341-858-7 **£25**

Guidance Note 5: Protection Against Electric Shock, 5th Edition
Order book PWG5170B 115pp Paperback 2008
ISBN: 978-0-86341-859-4 **£25**

Guidance Note 6: Protection Against Overcurrent, 5th Edition
Order book PWG6170B 113pp Paperback 2008
ISBN: 978-0-86341-860-0 **£25**

Guidance Note 7: Special Locations, 3rd Edition
Order book PWG7170B 142pp Paperback 2008
ISBN: 978-0-86341-861-7 **£25**

Guidance Note 8: Earthing & Bonding, 1st Edition
Order book PWRG0241 168pp Paperback 2007
ISBN: 978-0-86341-616-3 **£25**

continues overleaf ▶

Other guidance publications

**Commentary on IEE Wiring Regulations
(16th Edition, BS 7671:2001)**
Order book PBNS0310 438pp Hardback 2002
ISBN: 978-0-85296-237-4 £45
New edition due in 2008

Electrical Maintenance, 2nd Edition
Order book PWR05100 227pp Paperback 2006
ISBN: 978-0-86341-563-0 £35

**Electrical Plugs and Wiring and World
Electricity Supplies**
Order book PWR03070 272pp Paperback 2005
ISBN: 978-058045-158-4 £170

**Electrical Craft Principles, Volume 1,
4th Edition**
Order book PBST1010 318pp Paperback 1995
ISBN: 978-0-85296-811-6 £18
New edition due in 2008

**Electrical Craft Principles, Volume 2,
4th Edition**
Order book PBST1020 409pp Paperback 1995
ISBN: 978-0-85296-833-8 £18
New edition due in 2008

**Code of Practice for In-service Inspection
and Testing of Electrical Equipment,
3rd Edition**
Order book PWR08630 138pp Paperback 2007
ISBN: 978-0-86341-833-4 £35

**Code of Practice for Installation of
Electrical and Electronic Equipment
in Ships BS 8450**
Order book PWR03160 68pp Looseleaf 2006
ISBN: 978-0-58048-280-9 £70

**Recommendations for the Electrical and
Electronic Equipment of Mobile and Fixed
Offshore Installations**
Order book PWR04010 212pp Paperback 1992
ISBN: 978-0-85296-528-3 £60

1st Amendment
Order book PWR04020 28pp Booklet 1995
ISBN: 978-0-85296-846-8 £10

Electrical training courses

We offer a comprehensive range of technical training at many levels, serving your training and career development requirements as and when they arise.

Courses range from Electrical Basics to Qualifying City & Guilds or EAL awards.

Train to the 17th Edition BS 7671:2008
▶ Update from 16th to 17th Edition
▶ Understand the changes
▶ New qualifying awards C&G/EAL
▶ Meet industry standards

Qualifying courses
▶ Certificate of Competence Management of Electrical Equipment Maintenance (PAT) – 1 day
▶ Certificate of Competence for the Inspection and Testing of Electrical Equipment (PAT) – 1 day
▶ Certificate in the Requirements for Electrical Installations – 3 days
▶ Upgrade from 16th Edition achieved since 2001 – 1 day
▶ Certificate in Fundamental Inspection, Testing and Internal Verification – 3 days
▶ Certificate in Inspection, Testing and Certification of Electrical Installations – 3 days

Other 17th Edition courses
▶ Earthing & Bonding – For designers and electrical contractors who require a good working knowledge of the E & B arrangements as required by BS 7671:2008
▶ 17th Edition Design – BS 7671 and the principles associated with the design of electrical installations

To view all our current courses and book online, visit **www.theiet.org/coursesbr**

To discuss your training requirements and for on site group training, please speak to one of our advisors on +44 (0)1438 767289

For more information, visit www.theiet.org/wiringregs

ORDER FORM

Photocopies of this order form will be accepted.

How to order

BY PHONE:
+44 (0)1438 767328

BY FAX:
+44 (0)1438 767375

BY EMAIL:
sales@theiet.org

BY POST:
The Institution of Engineering and Technology,
PO Box 96,
Stevenage
SG1 2SD, UK

OVER THE WEB:
www.theiet.org/books

*Postage/handling: Please add £1.50/item for handling unless order is prepaid. Postage is free within the UK, outside UK add £7.50 per item (Europe) and £10.00 per item (Rest of World). Books will be sent via airmail. Courier rates are available on request.

**To qualify for discounts, member orders must be placed directly with the Institution.

GUARANTEED RIGHT OF RETURN:
If at all unsatisfied, you may return book(s) in new condition within 30 days for a full refund. Please include a copy of the invoice.

DATA PROTECTION:
The information that you provide to the IET will be used to ensure we provide you with products and services that best meet your needs. This may include the promotion of specific IET products and services by post and/or electronic means. By providing us with your email address and/or mobile telephone number you agree that we may contact you by electronic means. You can change this preference at any time by visiting www.theiet.org/my.

Details

Name:

Job Title:

Company/Institution:

Address:

Postcode: Country:

Tel: Fax:

Email:

Membership No (if Institution member):

Payment methods

☐ By **cheque** made payable to The Institution of Engineering and Technology
☐ By **credit/debit card:**
☐ Visa ☐ Mastercard ☐ American Express ☐ Maestro Issue No:_____ Valid from: ☐☐ ☐☐
Card No: ☐☐☐☐ ☐☐☐☐ ☐☐☐☐ ☐☐☐☐ Expiry Date: ☐☐ ☐☐
Signature_____ Card Security Code: ☐☐☐☐
(Orders not valid unless signed) (3 or 4 digits on reverse of card)

Cardholder Name:

Cardholder Address:

Postcode: Town:

Country:

☐ By official **company purchase order** (please attach copy)
EU VAT number:_____

Ordering information

Quantity	Book No.	Title/Author	Price (£)
		Subtotal	
		Member Discount**	
		+ Postage /Handling*	
		+ VAT (if applicable)	
		Total	

The Institution of Engineering and Technology (the IET) is a not for profit organisation, registered Charity No. 211014

Membership

Passionate about engineering? Committed to your career?

Do you want to join an organisation that is inspiring, insightful and innovative?

One of the most highly recognised knowledge sharing networks in the world, membership to the Institution of Engineering and Technology is for engineers and technologists working or studying in an increasingly multidisciplinary, digital and global environment.

Joining the IET and having access to tailored products and services will become invaluable for your career and can be your first step towards professional qualifications.

You could take advantage of ...

- Fortnightly copy of the industry's leading publication, *Engineering & Technology* magazine.
- Professional development and career support services to help gain registration.
- Dedicated training courses, seminars and events covering a wide range of subjects and skills.
- Watch live IET.tv event footage at your desktop via the internet, ask the speaker questions during live streaming and feel part of the audience without physically being there.
- Access to over 100 local networks around the world.
- Meet like-minded professionals through our array of specialist online communities.
- Instant online access to over 70,000 books, 3,000 periodicals and full-text collections of electronic articles – wherever you are in the world.
- Discounted rates on IET books and technical proceedings.

Join online today www.theiet.org/join or contact our membership and customer service centre on +44 (0)1438 765678

Professional Registration

What type of registration is for you?

Chartered Engineers (CEng) develop appropriate solutions to engineering problems, using new or existing technologies, through innovation, creativity and change. They might develop and apply new technologies, promote advanced designs and design methods, introduce new and more efficient production techniques, marketing and construction concepts, pioneer new engineering services and management methods. Chartered Engineers are engaged in technical and commercial leadership and possess interpersonal skills.

Incorporated Engineers (IEng) maintain and manage applications of current and developing technology, and may undertake engineering design, development, manufacture, construction and operation. Incorporated Engineers are engaged in technical and commercial management and possess effective interpersonal skills.

Engineering Technicians (EngTech) are involved in applying proven techniques and procedures to the solution of practical engineering problems. You will carry supervisory or technical responsibility, and are competent to exercise creative aptitudes and skills within defined fields of technology. Engineering Technicians also contribute to the design, development, manufacture, commissioning, operation or maintenance of products, equipment, processes or services.

For further information on Professional Registration (CEng/IEng/EngTech), tel: +44 (0)1438 767282 or email: membership@theiet.org

The Institution of *Engineering and Technology*

Application Form for Associates

PLEASE USE BLACK INK AND WRITE IN BLOCK CAPITALS. If you require assistance or advice, please telephone the Member and Customer Service Centre on **+44 (0)1438 765678** or visit **www.theiet.org/contactmembership**

PLEASE SEND THE COMPLETED FORM TO: **THE INSTITUTION OF ENGINEERING AND TECHNOLOGY, PO BOX 96, STEVENAGE, SG1 2SD, UK**
To apply online please visit **www.theiet.org/join**
Your subscription year will run annually from the month that you join the Institution. Join as an **Associate** for immediate access to all the IET products and services.

A: YOUR DETAILS

Title	First Name(s)

Surname/Family name

Date of Birth

Home Address

	Postcode

Tel No/Mobile

Email	Job Title

B: YOUR INTERESTS
Please select your area(s) of interest below. Tick all that apply.

☐ Communications ☐ Construction/Building Services ☐ Consumer Electronics ☐ Control ☐ Electronics ☐ IT
☐ Management ☐ Manufacturing ☐ Mechanical Engineering ☐ Power ☐ Transport

C: DECLARATION
I declare that the statements made on this form are to the best of my knowledge true. I confirm that I have not committed any offence of which the Institution of Engineering and Technology (IET) would require me to give notice under its Rules of Conduct. I agree to comply with the Royal Charter and Bye-laws of the Institution and whilst remaining a member I will do my best to promote the interests of the Institution. I understand and consent to the information provided on this form being processed by the Institution for its sole use and that of its associated organisations for the purposes of promoting, delivering and improving my experience of the Institution, its products and its services, by post and electronic means. The Rules of Conduct and the Royal Charter and Bye-laws are published on the website **www.theiet.org/byelaws**

Signature of Applicant	Date

D: YOUR PAYMENT OPTIONS

Subscription Rates from 1 January 2008 - 31st December 2008*	
Associates Standard Rate	£102
If your annual income from all sources is less than £18,000, or its equivalent in local currency, you can apply for a reduced subscription rate.**	
☐ I would like to apply for a reduced subscription rate	
Associates Reduced Rate	£51
** Please note that the Institution reserves the right to contact your employer to verify this claim.	

*For up-to-date rates please visit: **www.theiet.org/membershipfees**

If you are applying from China or India then please visit **www.theiet.org/membershipfees** to find out about the rates set in local currency of Renminbi (RMB) for **Associates** and **Members** in mainland China and Rupees for **Associates** and **Members** in India.

Please specify if you wish to apply for designatory letters ☐ **TMIET** ☐ **MIET**
And we will contact you for further information when processing your application

PUBGU2008

E: PAYMENT AMOUNT

1: Subscription fee £ ☐

2: I would like to donate ☐ £5 or £ ☐ to IET Connect, the IET Benevolent Fund. I am a UK taxpayer and would like IET Connect to reclaim the tax on this and all future payments under the Gift Aid Scheme.

3: Total £ ☐   IET Connect, the IET Benevolent Fund asks associates and members to donate at least £5 with their subscription each year. For more information on the services available visit www.ietconnect.org. IET Connect is a separate registered charity.

I wish to pay by: ☐ CARD   ☐ DIRECT DEBIT   ☐ CHEQUE/STERLING BANK DRAFT

Please complete the Payment By Card section below   Please complete the Direct Debit Instruction below   Please make this payable to The Institution of Engineering and Technology

F: PAYMENT DETAILS
Please complete one of the following options:

1: PAYMENT BY CARD
I authorise you to charge my Visa/Mastercard/American Express/Maestro/Delta* with: *Please delete as appropriate

☐ a one-off payment with the amount stated in Section E for my current year's subscription.

☐ the amount stated in Section E for my current year's subscription and payment of unspecified amounts for future years' subscriptions as and when they become due (Visa/Mastercard only). I understand that I will be given prior notice in writing.

Card Number
☐ ☐ ☐ ☐ ☐ ☐ ☐ ☐ ☐ ☐ ☐ ☐ ☐ ☐ ☐ ☐

Valid from [m] [m] [y] [y] ☐ ☐ ☐ ☐   Expiry date [m] [m] [y] [y] ☐ ☐ ☐ ☐   Issue Number ☐ ☐   3 or 4 digit security code (CSC) ☐ ☐ ☐ ☐

Name on credit card

Address

Postcode   Country

Signature   Date

Membership Number (Please complete if applicable)
☐ ☐ ☐ ☐ ☐ ☐ ☐ ☐ ☐

DIRECT Debit

2: INSTRUCTION TO YOUR BANK OR BUILDING SOCIETY TO PAY BY DIRECT DEBIT
Please fill in the whole form using a ball point pen and send to: The Institution of Engineering and Technology, PO Box 96, Stevenage, SG1 2SD, United Kingdom

To: The Manager Bank/Building Society

Address

Postcode

Name(s) of Account Holder(s)

Bank/Building Society Account Number   Branch Sort Code   Originator's Identification Number
☐ ☐ ☐ ☐ ☐ ☐ ☐ ☐   ☐ ☐ ☐ ☐ ☐ ☐   9 2 0 0 1 5

Reference (For office use only)

Instruction to your Bank or Building Society
Please pay the Institution Direct Debits from the account detailed in this instruction subject to the safeguards assured by the Direct Debit Guarantee. I understand that this instruction may remain with the Institution and, if so, details will be passed electronically to my Bank/Building Society.

Signature   Date

3: PAYMENT BY CHEQUE

☐ Enclosed is a cheque made payable to the Institution of Engineering and Technology. Please write your Name and Address on the reverse of the cheque.

This guarantee should be detached and retained by the Payer. The Direct Debit Guarantee
- This Guarantee is offered by all Banks and Building Societies that take part in the Direct Debit Scheme. The efficiency and security of the Scheme is monitored and protected by your own Bank or Building Society.
- If the amounts to be paid or the payment dates change the Institution of Engineering and Technology will notify you 14 working days in advance of your account being debited or as otherwise agreed.
- If an error is made by the Institution or your Bank or Building Society, you are guaranteed a full and immediate refund from your branch of the amount paid.
- You can cancel a Direct Debit at any time by writing to your Bank or Building Society. Please also send a copy of your letter to us.

Notes

Notes